Honda
TRX350, TRX250 and TRX250EX
ATV Owners Workshop Manual

**by Alan Ahlstrand
and John H Haynes**

Member of the Guild of Motoring Writers

Models covered:
TRX350 Rancher, 2000 through 2005
TRX250 Recon, 1997 through 2009
TRX250EX Sportrax, 2001 through 2009

(2553 - 9X2)

ABCDE
FGH

Haynes Publishing
Sparkford Nr Yeovil
Somerset BA22 7JJ England

Haynes North America, Inc
859 Lawrence Drive
Newbury Park
California 91320 USA
www.haynes.com

Acknowledgments

Our thanks to Craig Adams and Cal Coast Motorsports for providing the TXR350 machine used in the production of this manual. Thanks also to Grand Prix Sports, Santa Clara, California, for providing the TRX250; to Anthony Morter, service manager, for providing the shop facilities and arranging the teardown; and to Craig Wardner, service technician, for performing the teardown and providing valuable technical advice. Wiring diagrams produced exclusively for Haynes North America, Inc. by Ed McCahill.

A book in the Haynes Owners Workshop Manual Series

Printed in Malaysia

ISBN-13: 978-1-56392-778-2
ISBN-10: 1-56392-778-0

Library of Congress Control Number: 2009930744

Contents

Introductory pages

About this manual   0-5
Introduction to the Honda Rancher, Recon and Sportrax   0-5
Identification numbers   0-6
Buying parts   0-7
General specifications   0-8
Maintenance techniques, tools and working facilities   0-10
Safety first!   0-16
ATV chemicals and lubricants   0-17
Conversion factors   0-18
Fraction, decimal and millimeter equivalents   0-19
Troubleshooting   0-20

Chapter 1
Tune-up and routine maintenance   1-1

Chapter 2
Engine, clutch and transmission   2-1

Chapter 3
Fuel and exhaust systems   3-1

Chapter 4
Ignition system   4-1

Chapter 5
Steering, suspension and final drive   5-1

Chapter 6
Brakes, wheels and tires   6-1

Chapter 7
Bodywork and frame   7-1

Chapter 8
Electrical system   8-1

Wiring diagrams   8-23

Index   IND-1

Honda TRX350 Rancher (2004 TRX350 FE model)

About this manual

Its purpose

The purpose of this manual is to help you get the best value from your vehicle. It can do so in several ways. It can help you decide what work must be done, even if you choose to have it done by a dealer service department or a repair shop; it provides information and procedures for routine maintenance and servicing; and it offers diagnostic and repair procedures to follow when trouble occurs.

We hope you use the manual to tackle the work yourself. For many simpler jobs, doing it yourself may be quicker than arranging an appointment to get the vehicle into a shop and making the trips to leave it and pick it up. More importantly, a lot of money can be saved by avoiding the expense the shop must pass on to you to cover its labor and overhead costs. An added benefit is the sense of satisfaction and accomplishment that you feel after doing the job yourself.

Using the manual

The manual is divided into Chapters. Each Chapter is divided into numbered Sections, which are headed in bold type between horizontal lines. Each Section consists of consecutively numbered paragraphs.

At the beginning of each numbered Section you will be referred to any illustrations which apply to the procedures in that Section. The reference numbers used in illustration captions pinpoint the pertinent Section and the Step within that Section. That is, illustration 3.2 means the illustration refers to Section 3 and Step (or paragraph) 2 within that Section.

Procedures, once described in the text, are not normally repeated. When it's necessary to refer to another Chapter, the reference will be given as Chapter and Section number. Cross references given without use of the word "Chapter" apply to Sections and/or paragraphs in the same Chapter. For example, "see Section 8" means in the same Chapter.

References to the left or right side of the vehicle assume you are sitting on the seat, facing forward.

All-terrain vehicle manufacturers continually make changes to specifications and recommendations, and these, when notified, are incorporated into our manuals at the earliest opportunity.

Even though we have prepared this manual with extreme care, neither the publisher nor the author can accept responsibility for any errors in, or omissions from, the information given.

NOTE

A **Note** provides information necessary to properly complete a procedure or information which will make the procedure easier to understand.

CAUTION

A **Caution** provides a special procedure or special steps which must be taken while completing the procedure where the Caution is found. Not heeding a Caution can result in damage to the assembly being worked on.

WARNING

A **Warning** provides a special procedure or special steps which must be taken while completing the procedure where the Warning is found. Not heeding a Warning can result in personal injury.

Introduction to the Honda Rancher, Recon and Sportrax

The TRX350 Rancher is one of Honda's most popular utility all-terrain vehicles, occupying the middle of Honda's engine size range at 350cc. The TRX250 Recon is a utility ATV, and the TRX250EX is a sport version of the Recon.

All models use air-cooled pushrod engines with an auxiliary oil cooler. The crankshaft centerline is parallel to the centerline of the vehicle.

Fuel is delivered to the cylinder by a single carburetor.

The front suspension uses upper and lower control arms on each side of the vehicle, with a coil spring-shock absorber unit attached to each upper control arm.

The rear suspension on all models uses a swingarm with a single coil spring-shock absorber unit.

The Rancher and Recon use a sealed drum brake at each front wheel. The TRX250EX uses front disc brakes. All models use a single sealed drum brake at the rear.

Shaft final drive is used on all models covered in this manual. Four-wheel drive is optional on the Rancher.

Identification numbers

The frame serial number is stamped into the front of the frame and printed on a label affixed to the frame. The engine number is stamped into the right side of the crankcase. Both of these numbers should be recorded and kept in a safe place so they can be furnished to law enforcement officials in the event of a theft.

The frame serial number, engine serial number and carburetor identification number should also be kept in a handy place (such as with your driver's license) so they are always available when purchasing or ordering parts for your machine.

The models covered by this manual are as follows:

TRX350 Rancher, 2000 through 2005
TRX250 Recon, 1997 through 2009
TRX250EX Sportrax, 2001 through 2009

Identifying model years

The procedures in this manual identify the vehicles by model year. The model year is included in a decal on the frame, but in case the decal is missing or obscured, the following table identifies the initial frame number of each model year. **Note:** *Model year and frame number information is not available for 2006 and later models. However, the model year is printed on a sticker on the frame.*

350 models

The frame serial number is located at the front of the frame

The engine serial number is located on the crankcase

Year	Initial frame number
2000	
TRX350FE	478TE254*YA00001
TRX350FM.	478TE250*YA00001
TRX350TE	478TE244*YA00001
TRX350TM.	478TE240*YA00001
2001	
TRX350FE (A)	478TE254*1A100001
TRX350FE (2A)	478TE254*14000001
TRX350FM (A)	478TE250*1A100001
TRX350FM (2A)	478TE250*14000001
TRX350TE	478TE244*1A100001
TRX350TM.	478TE240*1A100001
2002	
TRX350FE (A)	478TE254*2A200001
TRX350FE (2A)	478TE254*24100001
TRX350FM (A)	478TE250*2A200001
TRX350FM (2A)	478TE250*24100001
TRX350TE (A)	478TE244*2A200001
TRX350TE (2A)	478TE244*24100001
TRX350TM (A)	478TE240*2A200001
TRX350TM (2A)	478TE240*24100001
2003	
TRX350FE (A)	478TE254*3A300001
TRX350FE (2A)	478TE254*34200001
TRX350FM (A)	478TE250*3A300001
TRX350FM (2A)	478TE250*34200001
TRX350TE (A)	478TE244*3A300001
TRX350TE (2A)	478TE244*34200001
TRX350TM (A)	478TE240*3A300001
TRX350TM (2A)	478TE240*34200001

350 models (continued)

Year	Initial frame number
2004	
TRX350FE (A)	478TE254*4A400001
TRX350FE (2A)	478TE254*44300001
TRX350FM (A)	478TE250*4A400001
TRX350FM (2A)	478TE250*44300001
TRX350TE (A)	478TE244*4A400001
TRX350TE (2A)	478TE244*44300001
TRX350TM (A)	478TE240*4A400001
TRX350TM (2A)	478TE240*44300001
2005	
TRX350FE (A)	478TE254*54400001
TRX350FM (A)	478TE250*54400001
TRX350TE (A)	478TE244*54400001
TRX350TM (A)	478TE240*54400001

250 Recon models

Year	Initial frame number
1997	
Except California	478TE210*VA00001
California	478TE211*VA00001
1998	
Except California	478TE210*WA000001
California	478TE211*WA000001

Year	Initial frame number
1999	478TE210*XA200001
2000	478TE210*Y400001
2001	478TE210*1410001
2002	
TRX250TE	478TE214*24200001
TRX250TM	478TE210*24200001
2003	
TRX250TE	478TE214*34300001
TRX250TM	478TE210*34300001
2004	
TRX250TE	478TE214*44400001
TRX250TM	478TE210*44400001
2005	
TRX250TE	478TE214*54500001
TRX250TM	478TE210*54500001

Sportrax 250EX models

Year	Initial frame number
2001	478TE270*1400001
2002	478TE270*2410001
2003	478TE270*3420001
2004	478TE270*4430001
2005	478TE270*5440001

Buying parts

Once you have found all the identification numbers, record them for reference when buying parts. Since the manufacturers change specifications, parts and vendors (companies that manufacture various components on the machine), providing the ID numbers is the only way to be reasonably sure that you are buying the correct parts.

Whenever possible, take the worn part to the dealer so direct comparison with the new component can be made. Along the trail from the manufacturer to the parts shelf, there are numerous places that the part can end up with the wrong number or be listed incorrectly.

The two places to purchase new parts for your vehicle - the accessory store and the franchised dealer - differ in the type of parts they carry. While dealers can obtain virtually every part for your vehicle, the accessory dealer is usually limited to normal high wear items such as shock absorbers, tune-up parts, various engine gaskets, cables, chains, brake parts, etc. Rarely will an accessory outlet have major suspension components, cylinders, transmission gears, or cases.

Used parts can be obtained for roughly half the price of new ones, but you can't always be sure of what you're getting. Once again, take your worn part to the wrecking yard (breaker) for direct comparison.

Whether buying new, used or rebuilt parts, the best course is to deal directly with someone who specializes in parts for your particular make.

General specifications

Rancher 350 models

Wheelbase
    2WD models .... 1253 mm (49.3 inches)
    4WD models.. 1246 mm (49.1 inches)
Overall length
    2000 through 2003 models .... 1983 mm (78.1 inches)
    2004 and later models .... 2031 mm (80.0 inches)
Overall width
    2000 through 2003 models .... 1143 mm (45.0 inches)
    2004 and later models .... 1114 mm (43.9 inches)
Overall height
    2000 through 2003 2WD models .... 1119 mm (44.1 inches)
    2000 through 2003 4WD models .... 1130 mm (44.5 inches)
    2004 and later 2WD models.................................. 1129 mm (44.4 inches)
    2004 and later 4WD models.................................. 1141 mm (44.9 inches)
Seat height
    2WD models.. 812 mm (32.0 inches)
    2000 through 2003 4WD models .... 824 mm (32.4 inches)
    2004 and later 4WD models.................................. 819 mm (32.3 inches)
Ground clearance
    2WD models.. 186 mm (7.3 inches)
    2000 through 2003 4WD models .... 184 mm (7.2 inches)
    2004 and later 4WD models.................................. Not specified
Dry weight
    2000 and 2001 2WD foot shift models .... 226 kg (498 lbs)
    2000 and 2001 2WD electric shift models .... 232 kg (511 lbs)
    2000 and 2001 4WD foot shift models...................... 237.5 kg (523.6 lbs)
    2000 and 2001 4WD electric shift models .... 242.5 kg (534.6 lbs)
    2002 and 2003 2WD foot shift models .... 226.5 kg (499.3 lbs)
    2002 and 2003 2WD electric shift models .... 232.5 kg (512.6 lbs)
    2002 and 2003 4WD foot shift models...................... 238 kg (525 lbs)
    2002 and 2003 4WD electric shift models .... 243 kg (536 lbs)
    2004 and later 2WD foot shift models .... 227 kg (500.0 lbs)
    2004 and later 2WD electric shift models .... 231 kg (509 lbs)
    2004 and later 4WD foot shift models...................... 238 kg (525 lbs)
    2004 and later 4WD electric shift models .... 241 kg (531 lbs)

Recon 250 models

Wheelbase
- 1997 through 2004 models .... 1131 mm (44.53 inches)
- 2005 through 2007 models .... 1130 mm (44.48 inches)
- 2008 and later models.. 1131 mm (44.53 inches)

Overall length
- 1997 through 2004 models .... 1794 mm (70.6 inches)
- 2005 and later models... 1905 mm (75.0 inches)

Overall width
- 1997 through 2004 models .... 1034 mm (40.71 inches)
- 2005 and later models... 1035 mm (40.74 inches)

Overall height
- 1997 through 2004 models .... 1054 mm (41.5 inches)
- 2005 through 2007 models .... 1065 mm (41.9 inches)
- 2008 and later models.. 1070 mm (42.1 inches)

Seat height
- 1997 through 2004 4WD models .... 777 mm (30.6 inches)
- 2005 through 2007 models .... 795 mm (31.3 inches)
- 2008 and later models.. 793 mm (31.2 inches)

Ground clearance
- 1997 through 2004 models .... 152 mm (6.0 inches)
- 2005 and later models... 150 mm (5.9 inches)

Sportrax 250EX models

Wheelbase.. 1124 mm (44.3 inches)

Overall length
- 2005 and earlier .... 1735 mm (68.3 inches)
- 2006 and later .... 1739 mm (68.5 inches)

Overall width.. 1062 mm (41.8 inches)

Overall height
- 2005 and earlier.. 1073 mm (42.2 inches)
- 2006 and 2007 .... 1076 mm (42.2 inches)
- 2008 and later .... 1082 mm (42.6 inches)

Ground clearance
- 2007 and earlier.. 149 mm (5.9 inches)
- 2008 and later .... 146 mm (5.7 inches)

Seat height
- 2005 and earlier.. 794 mm (31.3 inches)
- 2006 and later .... 797 mm (31.4 inches)

Maintenance techniques, tools and working facilities

Basic maintenance techniques

There are a number of techniques involved in maintenance and repair that will be referred to throughout this manual. Application of these techniques will enable the amateur mechanic to be more efficient, better organized and capable of performing the various tasks properly, which will ensure that the repair job is thorough and complete.

Fastening systems

Fasteners, basically, are nuts, bolts and screws used to hold two or more parts together. There are a few things to keep in mind when working with fasteners. Almost all of them use a locking device of some type (either a lock washer, locknut, locking tab or thread adhesive). All threaded fasteners should be clean, straight, have undamaged threads and undamaged corners on the hex head where the wrench fits. Develop the habit of replacing all damaged nuts and bolts with new ones.

Rusted nuts and bolts should be treated with a penetrating oil to ease removal and prevent breakage. Some mechanics use turpentine in a spout type oil can, which works quite well. After applying the rust penetrant, let it "work" for a few minutes before trying to loosen the nut or bolt. Badly rusted fasteners may have to be chiseled off or removed with a special nut breaker, available at tool stores.

If a bolt or stud breaks off in an assembly, it can be drilled out and removed with a special tool called an E-Z out (or screw extractor). Most dealer service departments and vehicle repair shops can perform this task, as well as others (such as the repair of threaded holes that have been stripped out).

Flat washers and lock washers, when removed from an assembly, should always be replaced exactly as removed. Replace any damaged washers with new ones. Always use a flat washer between a lock washer and any soft metal surface (such as aluminum), thin sheet metal or plastic. Special locknuts can only be used once or twice before they lose their locking ability and must be replaced.

Tightening sequences and procedures

When threaded fasteners are tightened, they are often tightened to a specific torque value (torque is basically a twisting force). Over-tightening the fastener can weaken it and cause it to break, while under-tightening can cause it to eventually come loose. Each bolt, depending on the material it's made of, the diameter of its shank and the material it is threaded into, has a specific torque value, which is noted in the Specifications. Be sure to follow the torque recommendations closely.

Fasteners laid out in a pattern (i.e. cylinder head bolts, engine case bolts, etc.) must be loosened or tightened in a sequence to avoid warping the component. Initially, the bolts/nuts should go on finger tight only. Next, they should be tightened one full turn each, in a criss-cross or diagonal pattern. After each one has been tightened one full turn, return to the first one tightened and tighten them all one half turn, following the same pattern. Finally, tighten each of them one quarter turn at a time until each fastener has been tightened to the proper torque. To loosen and remove the fasteners the procedure would be reversed.

Disassembly sequence

Component disassembly should be done with care and purpose to help ensure that the parts go back together properly during reassembly. Always keep track of the sequence in which parts are removed. Take note of special characteristics or marks on parts that can be installed more than one way (such as a grooved thrust washer on a shaft). It's a good idea to lay the disassembled parts out on a clean surface in the order that they were removed. It may also be helpful to make sketches or take instant photos of components before removal.

When removing fasteners from a component, keep track of their locations. Sometimes threading a bolt back in a part, or putting the washers and nut back on a stud, can prevent mixups later. If nuts and bolts can't be returned to their original locations, they should be kept in a compartmented box or a series of small boxes. A cupcake or muffin tin is ideal for this purpose, since each cavity can hold the bolts and nuts from a particular area (i.e. engine case bolts, valve cover bolts, engine mount bolts, etc.). A pan of this type is especially helpful when working on assemblies with very small parts (such as the carburetors and the valve train). The cavities can be marked with paint or tape to identify the contents.

Whenever wiring looms, harnesses or connectors are separated, it's a good idea to identify the two halves with numbered pieces of masking tape so they can be easily reconnected.

Gasket sealing surfaces

Throughout any vehicle, gaskets are used to seal the mating surfaces between components and keep lubricants, fluids, vacuum or pressure contained in an assembly.

Many times these gaskets are coated with a liquid or paste type gasket sealing compound before assembly. Age, heat and pressure can sometimes cause the two parts to stick together so tightly that they are very difficult to separate. In most cases, the part can be

Spark plug gap adjusting tool

Feeler gauge set

loosened by striking it with a soft-faced hammer near the mating surfaces. A regular hammer can be used if a block of wood is placed between the hammer and the part. Do not hammer on cast parts or parts that could be easily damaged. With any particularly stubborn part, always recheck to make sure that every fastener has been removed.

Avoid using a screwdriver or bar to pry apart components, as they can easily mar the gasket sealing surfaces of the parts (which must remain smooth). If prying is absolutely necessary, use a piece of wood, but keep in mind that extra clean-up will be necessary if the wood splinters.

After the parts are separated, the old gasket must be carefully scraped off and the gasket surfaces cleaned. Stubborn gasket material can be soaked with a gasket remover (available in aerosol cans) to soften it so it can be easily scraped off. A scraper can be fashioned from a piece of copper tubing by flattening and sharpening one end. Copper is recommended because it is usually softer than the surfaces to be scraped, which reduces the chance of gouging the part. Some gaskets can be removed with a wire brush, but regardless of the method used, the mating surfaces must be left clean and smooth. If for some reason the gasket surface is gouged, then a gasket sealer thick enough to fill scratches will have to be used during reassembly of the components. For most applications, a non-drying (or semi-drying) gasket sealer is best.

Hose removal tips

Hose removal precautions closely parallel gasket removal precautions. Avoid scratching or gouging the surface that the hose mates against or the connection may leak. Because of various chemical reactions, the rubber in hoses can bond itself to the metal spigot that the hose fits over. To remove a hose, first loosen the hose clamps that secure it to the spigot. Then, with slip joint pliers, grab the hose at the clamp and rotate it around the spigot. Work it back and forth until it is completely free, then pull it off (silicone or other lubricants will ease removal if they can be applied between the hose and the outside of the spigot). Apply the same lubricant to the inside of the hose and the outside of the spigot to simplify installation.

If a hose clamp is broken or damaged, do not reuse it. Also, do not reuse hoses that are cracked, split or torn.

Tools

A selection of good tools is a basic requirement for anyone who plans to maintain and repair a vehicle. For the owner who has few tools, if any, the initial investment might seem high, but when compared to the spiraling costs of routine maintenance and repair, it is a wise one.

Control cable pressure luber

To help the owner decide which tools are needed to perform the tasks detailed in this manual, the following tool lists are offered: Maintenance and minor repair, Repair and overhaul and Special. The newcomer to practical mechanics should start off with the Maintenance and minor repair tool kit, which is adequate for the simpler jobs. Then, as confidence and experience grow, the owner can tackle more difficult tasks, buying additional tools as they are needed. Eventually the basic kit will be built into the Repair and overhaul tool set. Over a period of time, the experienced do-it-yourselfer will assemble a tool set complete enough for most repair and overhaul procedures and will add tools from the Special category when it is felt that the expense is justified by the frequency of use.

Maintenance and minor repair tool kit

The tools in this list should be considered the minimum required for performance of routine maintenance, servicing and minor repair work. We recommend the purchase of combination wrenches (box end and open end combined in one wrench); while more expensive than open-ended ones, they offer the advantages of both types of wrench.

Combination wrench set (6 mm to 22 mm)
Adjustable wrench - 8 in
Spark plug socket (with rubber insert)

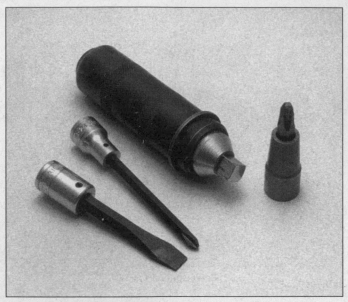

Hand impact screwdriver and bits

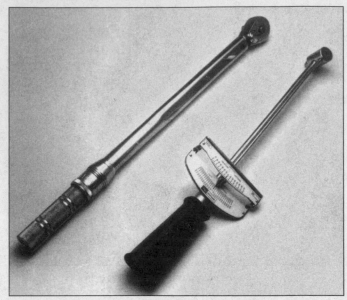

Torque wrenches (left - click type; right, beam type)

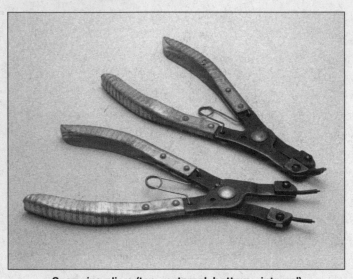

Snap-ring pliers (top - external; bottom - internal)

Allen wrenches (left), and Allen head sockets (right)

Spark plug gap adjusting tool
Feeler gauge set
Standard screwdriver (5/16 in x 6 in)
Phillips screwdriver (No. 2 x 6 in)
Allen (hex) wrench set (4 mm to 12 mm)
Combination (slip-joint) pliers - 6 in
Hacksaw and assortment of blades
Tire pressure gauge
Control cable pressure luber
Grease gun
Oil can
Fine emery cloth
Wire brush
Hand impact screwdriver and bits
Funnel (medium size)
Safety goggles
Drain pan
Work light with extension cord

Repair and overhaul tool set

These tools are essential for anyone who plans to perform major repairs and are intended to supplement those in the Maintenance and minor repair tool kit. Included is a comprehensive set of sockets which, though expensive, are invaluable because of their versatility (especially when various extensions and drives are available). We recommend the 3/8 inch drive over the 1/2 inch drive for general vehicle maintenance and repair (ideally, the mechanic would have a 3/8 inch drive set and a 1/2 inch drive set).

Alternator rotor puller tool
Socket set(s)
Reversible ratchet
Extension - 6 in
Universal joint
Torque wrench (same size drive as sockets)
Ball peen hammer - 8 oz
Soft-faced hammer (plastic/rubber)
Standard screwdriver (1/4 in x 6 in)

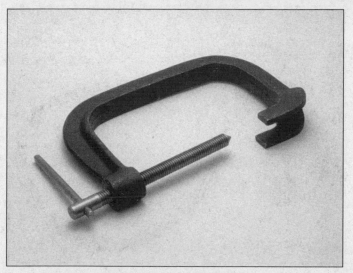

Valve spring compressor

Piston ring removal/installation tool

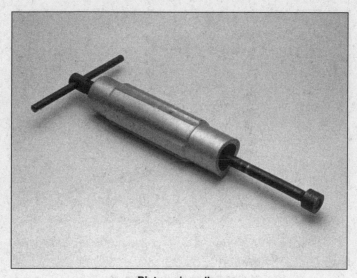

Piston pin puller

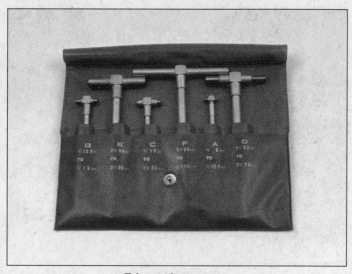

Telescoping gauges

Standard screwdriver (stubby - 5/16 in)
Phillips screwdriver (No. 3 x 8 in)
Phillips screwdriver (stubby - No. 2)
Pliers - locking
Pliers - lineman's
Pliers - needle nose
Pliers - snap-ring (internal and external)
Cold chisel - 1/2 in
Scriber
Scraper (made from flattened copper tubing)
Center punch
Pin punches (1/16, 1/8, 3/16 in)
Steel rule/straightedge - 12 in
Pin-type spanner wrench
A selection of files
Wire brush (large)

Note: Another tool which is often useful is an electric drill with a chuck capacity of 3/8 inch (and a set of good quality drill bits).

Special tools

The tools in this list include those which are not used regularly, are expensive to buy, or which need to be used in accordance with their

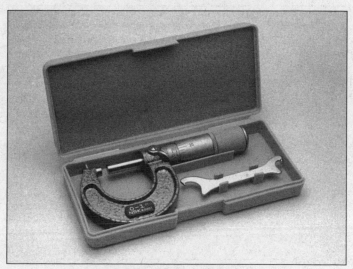

0-to-1 inch micrometer

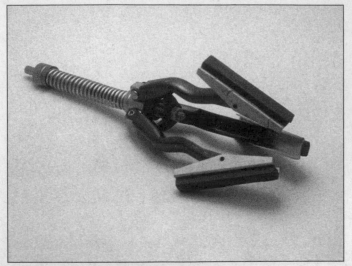

Cylinder surfacing hone

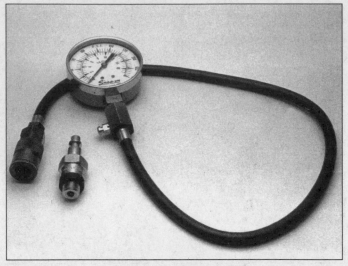

Cylinder compression gauge

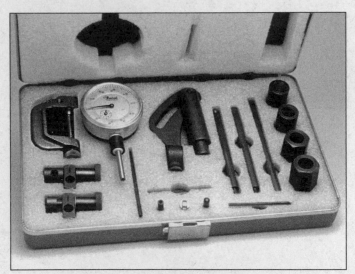

Dial indicator set

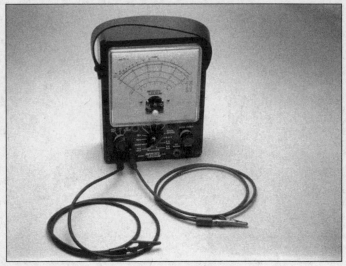

Multimeter (volt/ohm/ammeter)

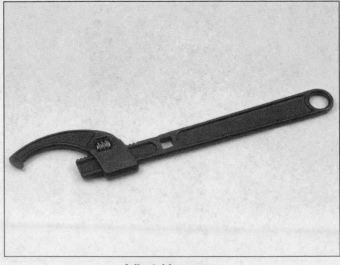

Adjustable spanner

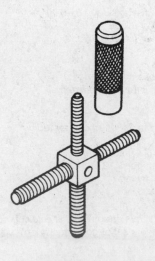

Alternator rotor puller

manufacturer's instructions. Unless these tools will be used frequently, it is not very economical to purchase many of them. A consideration would be to split the cost and use between yourself and a friend or friends (i.e. members of an ATV club).

This list primarily contains tools and instruments widely available to the public, as well as some special tools produced by the vehicle manufacturer for distribution to dealer service departments. As a result, references to the manufacturer's special tools are occasionally included in the text of this manual. Generally, an alternative method of doing the job without the special tool is offered. However, sometimes there is no alternative to their use. Where this is the case, and the tool can't be purchased or borrowed, the work should be turned over to the dealer service department or a vehicle repair shop.

Valve spring compressor
Piston ring removal and installation tool
Piston pin puller
Telescoping gauges
Micrometer(s) and/or dial/Vernier calipers
Cylinder surfacing hone
Cylinder compression gauge
Dial indicator set
Multimeter
Adjustable spanner
Manometer or vacuum gauge set
Small air compressor with blow gun and tire chuck

Buying tools

For the do-it-yourselfer who is just starting to get involved in vehicle maintenance and repair, there are a number of options available when purchasing tools. If maintenance and minor repair is the extent of the work to be done, the purchase of individual tools is satisfactory. If, on the other hand, extensive work is planned, it would be a good idea to purchase a modest tool set from one of the large retail chain stores. A set can usually be bought at a substantial savings over the individual tool prices (and they often come with a tool box). As additional tools are needed, add-on sets, individual tools and a larger tool box can be purchased to expand the tool selection. Building a tool set gradually allows the cost of the tools to be spread over a longer period of time and gives the mechanic the freedom to choose only those tools that will actually be used.

Tool stores and vehicle dealers will often be the only source of some of the special tools that are needed, but regardless of where tools are bought, try to avoid cheap ones (especially when buying screwdrivers and sockets) because they won't last very long. There are plenty of tools around at reasonable prices, but always aim to purchase items which meet the relevant national safety standards. The expense involved in replacing cheap tools will eventually be greater than the initial cost of quality tools.

It is obviously not possible to cover the subject of tools fully here.

For those who wish to learn more about tools and their use, there is a book entitled Motorcycle Workshop Practice Manual (Book no. 1454) available from the publishers of this manual. It also provides an introduction to basic workshop practice which will be of interest to a home mechanic working on any type of vehicle.

Care and maintenance of tools

Good tools are expensive, so it makes sense to treat them with respect. Keep them clean and in usable condition and store them properly when not in use. Always wipe off any dirt, grease or metal chips before putting them away. Never leave tools lying around in the work area.

Some tools, such as screwdrivers, pliers, wrenches and sockets, can be hung on a panel mounted on the garage or workshop wall, while others should be kept in a tool box or tray. Measuring instruments, gauges, meters, etc. must be carefully stored where they can't be damaged by weather or impact from other tools.

When tools are used with care and stored properly, they will last a very long time. Even with the best of care, tools will wear out if used frequently. When a tool is damaged or worn out, replace it; subsequent jobs will be safer and more enjoyable if you do.

Working facilities

Not to be overlooked when discussing tools is the workshop. If anything more than routine maintenance is to be carried out, some sort of suitable work area is essential.

It is understood, and appreciated, that many home mechanics do not have a good workshop or garage available and end up removing an engine or doing major repairs outside (it is recommended, however, that the overhaul or repair be completed under the cover of a roof).

A clean, flat workbench or table of comfortable working height is an absolute necessity. The workbench should be equipped with a vise that has a jaw opening of at least four inches.

As mentioned previously, some clean, dry storage space is also required for tools, as well as the lubricants, fluids, cleaning solvents, etc. which soon become necessary.

Sometimes waste oil and fluids, drained from the engine or cooling system during normal maintenance or repairs, present a disposal problem. To avoid pouring them on the ground or into a sewage system, simply pour the used fluids into large containers, seal them with caps and take them to an authorized disposal site or service station. Plastic jugs are ideal for this purpose.

Always keep a supply of old newspapers and clean rags available. Old towels are excellent for mopping up spills. Many mechanics use rolls of paper towels for most work because they are readily available and disposable. To help keep the area under the vehicle clean, a large cardboard box can be cut open and flattened to protect the garage or shop floor.

Whenever working over a painted surface (such as the fuel tank) cover it with an old blanket or bedspread to protect the finish.

Safety first

Regardless of how enthusiastic you may be about getting on with the job at hand, take the time to ensure that your safety is not jeopardized. A moment's lack of attention can result in an accident, as can failure to observe certain simple safety precautions. The possibility of an accident will always exist, and the following points should not be considered a comprehensive list of all dangers. Rather, they are intended to make you aware of the risks and to encourage a safety conscious approach to all work you carry out on your vehicle.

Essential DOs and DON'Ts

DON'T rely on a jack when working under the vehicle. Always use approved jackstands to support the weight of the vehicle and place them under the recommended lift or support points.

DON'T attempt to loosen extremely tight fasteners (i.e. wheel lug nuts) while the vehicle is on a jack - it may fall.

DON'T start the engine without first making sure that the transmission is in Neutral (or Park where applicable) and the parking brake is set.

DON'T remove the radiator cap from a hot cooling system - let it cool or cover it with a cloth and release the pressure gradually.

DON'T attempt to drain the engine oil until you are sure it has cooled to the point that it will not burn you.

DON'T touch any part of the engine or exhaust system until it has cooled sufficiently to avoid burns.

DON'T siphon toxic liquids such as gasoline, antifreeze and brake fluid by mouth, or allow them to remain on your skin.

DON'T inhale brake lining dust - it is potentially hazardous (see *Asbestos* below).

DON'T allow spilled oil or grease to remain on the floor - wipe it up before someone slips on it.

DON'T use loose fitting wrenches or other tools which may slip and cause injury.

DON'T push on wrenches when loosening or tightening nuts or bolts. Always try to pull the wrench toward you. If the situation calls for pushing the wrench away, push with an open hand to avoid scraped knuckles if the wrench should slip.

DON'T attempt to lift a heavy component alone - get someone to help you.

DON'T *rush or take unsafe shortcuts to finish a job.*

DON'T allow children or animals in or around the vehicle while you are working on it.

DO wear eye protection when using power tools such as a drill, sander, bench grinder, etc. and when working under a vehicle.

DO keep loose clothing and long hair well out of the way of moving parts.

DO make sure that any hoist used has a safe working load rating adequate for the job.

DO get someone to check on you periodically when working alone on a vehicle.

DO carry out work in a logical sequence and make sure that everything is correctly assembled and tightened.

DO keep chemicals and fluids tightly capped and out of the reach of children and pets.

DO remember that your vehicle's safety affects that of yourself and others. If in doubt on any point, get professional advice.

Steering, suspension and brakes

These systems are essential to driving safety, so make sure you have a qualified shop or individual check your work. Also, compressed suspension springs can cause injury if released suddenly - be sure to use a spring compressor.

Airbags

Airbags are explosive devices that can CAUSE injury if they deploy while you're working on the vehicle. Follow the manufacturer's instructions to disable the airbag whenever you're working in the vicinity of airbag components.

Asbestos

Certain friction, insulating, sealing, and other products - such as brake linings, brake bands, clutch linings, torque converters, gaskets, etc. - may contain asbestos or other hazardous friction material. Extreme care must be taken to avoid inhalation of dust from such products, since it is hazardous to health. If in doubt, assume that they do contain asbestos.

Fire

Remember at all times that gasoline is highly flammable. Never smoke or have any kind of open flame around when working on a vehicle. But the risk does not end there. A spark caused by an electrical short circuit, by two metal surfaces contacting each other, or even by static electricity built up in your body under certain conditions, can ignite gasoline vapors, which in a confined space are highly explosive. Do not, under any circumstances, use gasoline for cleaning parts. Use an approved safety solvent.

Always disconnect the battery ground (-) cable at the battery before working on any part of the fuel system or electrical system. Never risk spilling fuel on a hot engine or exhaust component. It is strongly recommended that a fire extinguisher suitable for use on fuel and electrical fires be kept handy in the garage or workshop at all times. Never try to extinguish a fuel or electrical fire with water.

Fumes

Certain fumes are highly toxic and can quickly cause unconsciousness and even death if inhaled to any extent. Gasoline vapor falls into this category, as do the vapors from some cleaning solvents. Any draining or pouring of such volatile fluids should be done in a well ventilated area.

When using cleaning fluids and solvents, read the instructions on the container carefully. Never use materials from unmarked containers.

Never run the engine in an enclosed space, such as a garage. Exhaust fumes contain carbon monoxide, which is extremely poisonous. If you need to run the engine, always do so in the open air, or at least have the rear of the vehicle outside the work area.

The battery

Never create a spark or allow a bare light bulb near a battery. They normally give off a certain amount of hydrogen gas, which is highly explosive.

Always disconnect the battery ground (-) cable at the battery before working on the fuel or electrical systems.

If possible, loosen the filler caps or cover when charging the battery from an external source (this does not apply to sealed or maintenance-free batteries). Do not charge at an excessive rate or the battery may burst.

Take care when adding water to a non maintenance-free battery and when carrying a battery. The electrolyte, even when diluted, is very corrosive and should not be allowed to contact clothing or skin.

Always wear eye protection when cleaning the battery to prevent the caustic deposits from entering your eyes.

Household current

When using an electric power tool, inspection light, etc., which operates on household current, always make sure that the tool is correctly connected to its plug and that, where necessary, it is properly grounded. Do not use such items in damp conditions and, again, do not create a spark or apply excessive heat in the vicinity of fuel or fuel vapor.

Secondary ignition system voltage

A severe electric shock can result from touching certain parts of the ignition system (such as the spark plug wires) when the engine is running or being cranked, particularly if components are damp or the insulation is defective. In the case of an electronic ignition system, the secondary system voltage is much higher and could prove fatal.

Hydrofluoric acid

This extremely corrosive acid is formed when certain types of synthetic rubber, found in some O-rings, oil seals, fuel hoses, etc. are exposed to temperatures above 750-degrees F (400-degrees C). The rubber changes into a charred or sticky substance containing the acid. Once formed, the acid remains dangerous for years. If it gets onto the skin, it may be necessary to amputate the limb concerned.

When dealing with a vehicle which has suffered a fire, or with components salvaged from such a vehicle, wear protective gloves and discard them after use.

ATV chemicals and lubricants

A number of chemicals and lubricants are available for use in motorcycle maintenance and repair. They include a wide variety of products ranging from cleaning solvents and degreasers to lubricants and protective sprays for rubber, plastic and vinyl.

Contact point/spark plug cleaner is a solvent used to clean oily film and dirt from points, grime from electrical connectors and oil deposits from spark plugs. It is oil free and leaves no residue. It can also be used to remove gum and varnish from carburetor jets and other orifices.

Carburetor cleaner is similar to contact point/spark plug cleaner but it usually has a stronger solvent and may leave a slight oily residue. It is not recommended for cleaning electrical components or connections.

Brake system cleaner is used to remove brake dust, grease and brake fluid from the brake system, where clean surfaces are absolutely necessary. It leaves no residue and often eliminates brake squeal caused by contaminants.

Silicone-based lubricants are used to protect rubber parts such as hoses and grommets, and are used as lubricants for hinges and locks.

Multi-purpose grease is an all purpose lubricant used wherever grease is more practical than a liquid lubricant such as oil. Some multi-purpose grease is colored white and specially formulated to be more resistant to water than ordinary grease.

Gear oil (sometimes called gear lube) is a specially designed oil used in transmissions and final drive units, as well as other areas where high friction, high temperature lubrication is required. It is available in a number of viscosities (weights) for various applications.

Motor oil is the lubricant formulated for use in engines. It normally contains a wide variety of additives to prevent corrosion and reduce foaming and wear. Motor oil comes in various weights (viscosity ratings) from 0 to 50. The recommended weight of the oil depends on the season, temperature and the demands on the engine. Light oil is used in cold climates and under light load conditions. Heavy oil is used in hot climates and where high loads are encountered. Multi-viscosity oils are designed to have characteristics of both light and heavy oils and are available in a number of weights from 0W-20 to 20W-50.

Gasoline additives perform several functions, depending on their chemical makeup. They usually contain solvents that help dissolve gum and varnish that build up on carburetor and inlet parts. They also serve to break down carbon deposits that form on the inside surfaces of the combustion chambers. Some additives contain upper cylinder lubricants for valves and piston rings.

Brake and clutch fluid is a specially formulated hydraulic fluid that can withstand the heat and pressure encountered in break/clutch systems. Care must be taken that this fluid does not come in contact with painted surfaces or plastics. An opened container should always be resealed to prevent contamination by water or dirt.

Chain lubricants are formulated especially for use on motorcycle final drive chains. A good chain lube should adhere well and have good penetrating qualities to be effective as a lubricant inside the chain and on the side plates, pins and rollers. Most chain lubes are either the foaming type or quick drying type and are usually marketed as sprays. Take care to use a lubricant marked as being suitable for O-ring chains.

Degreasers are heavy duty solvents used to remove grease and grime that may accumulate on the engine and frame components. They can be sprayed or brushed on and, depending on the type, are rinsed with either water or solvent.

Solvents are used alone or in combination with degreasers to clean parts and assemblies during repair and overhaul. The home mechanic should use only solvents that are non-flammable and that do not produce irritating fumes.

Gasket sealing compounds may be used in conjunction with gaskets, to improve their sealing capabilities, or alone, to seal metal-to-metal joints. Many gasket sealers can withstand extreme heat, some are impervious to gasoline and lubricants, while others are capable of filling and sealing large cavities. Depending on the intended use, gasket sealers either dry hard or stay relatively soft and pliable. They are usually applied by hand, with a brush or are sprayed on the gasket sealing surfaces.

Thread locking compound is an adhesive locking compound that prevents threaded fasteners from loosening because of vibration. It is available in a variety of types for different applications.

Moisture dispersants are usually sprays that can be used to dry out electrical components such as the fuse block and wiring connectors. Some types an also be used as treatment for rubber and as a lubricant for hinges, cables and locks.

Waxes and polishes are used to help protect painted and plated surfaces from the weather. Different types of pain may require the use of different types of wax polish. Some polishes utilize a chemical or abrasive cleaner to help remove the top layer of oxidized (dull) paint on older vehicles. In recent years, many non-wax polishes (that contain a wide variety of chemicals such as polymers and silicones) have been introduced. These non-wax polishes are usually easier to apply and last longer than conventional waxes and polishes.

Conversion factors

Length (distance)

Inches (in)	X	25.4	= Millimeters (mm)	X 0.0394	= Inches (in)
Feet (ft)	X	0.305	= Meters (m)	X 3.281	= Feet (ft)
Miles	X	1.609	= Kilometers (km)	X 0.621	= Miles

Volume (capacity)

Cubic inches (cu in; in^3)	X	16.387	= Cubic centimeters (cc; cm^3)	X 0.061	= Cubic inches (cu in; in^3)
Imperial pints (Imp pt)	X	0.568	= Liters (l)	X 1.76	= Imperial pints (Imp pt)
Imperial quarts (Imp qt)	X	1.137	= Liters (l)	X 0.88	= Imperial quarts (Imp qt)
Imperial quarts (Imp qt)	X	1.201	= US quarts (US qt)	X 0.833	= Imperial quarts (Imp qt)
US quarts (US qt)	X	0.946	= Liters (l)	X 1.057	= US quarts (US qt)
Imperial gallons (Imp gal)	X	4.546	= Liters (l)	X 0.22	= Imperial gallons (Imp gal)
Imperial gallons (Imp gal)	X	1.201	= US gallons (US gal)	X 0.833	= Imperial gallons (Imp gal)
US gallons (US gal)	X	3.785	= Liters (l)	X 0.264	= US gallons (US gal)

Mass (weight)

Ounces (oz)	X	28.35	= Grams (g)	X 0.035	= Ounces (oz)
Pounds (lb)	X	0.454	= Kilograms (kg)	X 2.205	= Pounds (lb)

Force

Ounces-force (ozf; oz)	X	0.278	= Newtons (N)	X 3.6	= Ounces-force (ozf; oz)
Pounds-force (lbf; lb)	X	4.448	= Newtons (N)	X 0.225	= Pounds-force (lbf; lb)
Newtons (N)	X	0.1	= Kilograms-force (kgf; kg)	X 9.81	= Newtons (N)

Pressure

Pounds-force per square inch (psi; lbf/in^2; lb/in^2)	X	0.070	= Kilograms-force per square centimeter (kgf/cm^2; kg/cm^2)	X 14.223	= Pounds-force per square inch (psi; lbf/in^2; lb/in^2)
Pounds-force per square inch (psi; lbf/in^2; lb/in^2)	X	0.068	= Atmospheres (atm)	X 14.696	= Pounds-force per square inch (psi; lbf/in^2; lb/in^2)
Pounds-force per square inch (psi; lbf/in^2; lb/in^2)	X	0.069	= Bars	X 14.5	= Pounds-force per square inch (psi; lbf/in^2; lb/in^2)
Pounds-force per square inch (psi; lbf/in^2; lb/in^2)	X	6.895	= Kilopascals (kPa)	X 0.145	= Pounds-force per square inch (psi; lbf/in^2; lb/in^2)
Kilopascals (kPa)	X	0.01	= Kilograms-force per square centimeter (kgf/cm^2; kg/cm^2)	X 98.1	= Kilopascals (kPa)

Torque (moment of force)

Pounds-force inches (lbf in; lb in)	X	1.152	= Kilograms-force centimeter (kgf cm; kg cm)	X 0.868	= Pounds-force inches (lbf in; lb in)
Pounds-force inches (lbf in; lb in)	X	0.113	= Newton meters (Nm)	X 8.85	= Pounds-force inches (lbf in; lb in)
Pounds-force inches (lbf in; lb in)	X	0.083	= Pounds-force feet (lbf ft; lb ft)	X 12	= Pounds-force inches (lbf in; lb in)
Pounds-force feet (lbf ft; lb ft)	X	0.138	= Kilograms-force meters (kgf m; kg m)	X 7.233	= Pounds-force feet (lbf ft; lb ft)
Pounds-force feet (lbf ft; lb ft)	X	1.356	= Newton meters (Nm)	X 0.738	= Pounds-force feet (lbf ft; lb ft)
Newton meters (Nm)	X	0.102	= Kilograms-force meters (kgf m; kg m)	X 9.804	= Newton meters (Nm)

Vacuum

Inches mercury (in. Hg)	X	3.377	= Kilopascals (kPa)	X 0.2961	= Inches mercury
Inches mercury (in. Hg)	X	25.4	= Millimeters mercury (mm Hg)	X 0.0394	= Inches mercury

Power

Horsepower (hp)	X	745.7	= Watts (W)	X 0.0013	= Horsepower (hp)

Velocity (speed)

Miles per hour (miles/hr; mph)	X	1.609	= Kilometers per hour (km/hr; kph)	X 0.621	= Miles per hour (miles/hr; mph)

Fuel consumption*

Miles per gallon, Imperial (mpg)	X	0.354	= Kilometers per liter (km/l)	X 2.825	= Miles per gallon, Imperial (mpg)
Miles per gallon, US (mpg)	X	0.425	= Kilometers per liter (km/l)	X 2.352	= Miles per gallon, US (mpg)

Temperature

Degrees Fahrenheit = (°C x 1.8) + 32   Degrees Celsius (Degrees Centigrade; °C) = (°F - 32) x 0.56

*It is common practice to convert from miles per gallon (mpg) to liters/100 kilometers (l/100km),
where mpg (Imperial) x l/100 km = 282 and mpg (US) x l/100 km = 235

DECIMALS to MILLIMETERS

Decimal	mm	Decimal	mm
0.001	0.0254	0.500	12.7000
0.002	0.0508	0.510	12.9540
0.003	0.0762	0.520	13.2080
0.004	0.1016	0.530	13.4620
0.005	0.1270	0.540	13.7160
0.006	0.1524	0.550	13.9700
0.007	0.1778	0.560	14.2240
0.008	0.2032	0.570	14.4780
0.009	0.2286	0.580	14.7320
0.010	0.2540	0.590	14.9860
0.020	0.5080	0.600	15.2400
0.030	0.7620	0.610	15.4940
0.040	1.0160	0.620	15.7480
0.050	1.2700	0.630	16.0020
0.060	1.5240	0.640	16.2560
0.070	1.7780	0.650	16.5100
0.080	2.0320	0.660	16.7640
0.090	2.2860	0.670	17.0180
0.100	2.5400	0.680	17.2720
0.110	2.7940	0.690	17.5260
0.120	3.0480	0.700	17.7800
0.130	3.3020	0.710	18.0340
0.140	3.5560	0.720	18.2880
0.150	3.8100	0.730	18.5420
0.160	4.0640	0.740	18.7960
0.170	4.3180	0.750	19.0500
0.180	4.5720	0.760	19.3040
0.190	4.8260	0.770	19.5580
0.200	5.0800	0.780	19.8120
0.210	5.3340	0.790	20.0660
0.220	5.5880	0.800	20.3200
0.230	5.8420	0.810	20.5740
0.240	6.0960	0.820	20.8280
0.250	6.3500	0.830	21.0820
0.260	6.6040	0.840	21.3360
0.270	6.8580	0.850	21.5900
0.280	7.1120	0.860	21.8440
0.290	7.3660	0.870	22.0980
0.300	7.6200	0.880	22.3520
0.310	7.8740	0.890	22.6060
0.320	8.1280	0.900	22.8600
0.330	8.3820	0.910	23.1140
0.340	8.6360	0.920	23.3680
0.350	8.8900	0.930	23.6220
0.360	9.1440	0.940	23.8760
0.370	9.3980	0.950	24.1300
0.380	9.6520	0.960	24.3840
0.390	9.9060	0.970	24.6380
0.400	10.1600	0.980	24.8920
0.410	10.4140	0.990	25.1460
0.420	10.6680	1.000	25.4000
0.430	10.9220		
0.440	11.1760		
0.450	11.4300		
0.460	11.6840		
0.470	11.9380		
0.480	12.1920		
0.490	12.4460		

FRACTIONS to DECIMALS to MILLIMETERS

Fraction	Decimal	mm	Fraction	Decimal	mm
1/64	0.0156	0.3969	33/64	0.5156	13.0969
1/32	0.0312	0.7938	17/32	0.5312	13.4938
3/64	0.0469	1.1906	35/64	0.5469	13.8906
1/16	0.0625	1.5875	9/16	0.5625	14.2875
5/64	0.0781	1.9844	37/64	0.5781	14.6844
3/32	0.0938	2.3812	19/32	0.5938	15.0812
7/64	0.1094	2.7781	39/64	0.6094	15.4781
1/8	0.1250	3.1750	5/8	0.6250	15.8750
9/64	0.1406	3.5719	41/64	0.6406	16.2719
5/32	0.1562	3.9688	21/32	0.6562	16.6688
11/64	0.1719	4.3656	43/64	0.6719	17.0656
3/16	0.1875	4.7625	11/16	0.6875	17.4625
13/64	0.2031	5.1594	45/64	0.7031	17.8594
7/32	0.2188	5.5562	23/32	0.7188	18.2562
15/64	0.2344	5.9531	47/64	0.7344	18.6531
1/4	0.2500	6.3500	3/4	0.7500	19.0500
17/64	0.2656	6.7469	49/64	0.7656	19.4469
9/32	0.2812	7.1438	25/32	0.7812	19.8438
19/64	0.2969	7.5406	51/64	0.7969	20.2406
5/16	0.3125	7.9375	13/16	0.8125	20.6375
21/64	0.3281	8.3344	53/64	0.8281	21.0344
11/32	0.3438	8.7312	27/32	0.8438	21.4312
23/64	0.3594	9.1281	55/64	0.8594	21.8281
3/8	0.3750	9.5250	7/8	0.8750	22.2250
25/64	0.3906	9.9219	57/64	0.8906	22.6219
13/32	0.4062	10.3188	29/32	0.9062	23.0188
27/64	0.4219	10.7156	59/64	0.9219	23.4156
7/16	0.4375	11.1125	15/16	0.9375	23.8125
29/64	0.4531	11.5094	61/64	0.9531	24.2094
15/32	0.4688	11.9062	31/32	0.9688	24.6062
31/64	0.4844	12.3031	63/64	0.9844	25.0031
1/2	0.5000	12.7000	1	1.0000	25.4000

Troubleshooting

Contents

Symptom	Section

Engine doesn't start or is difficult to start
Starter motor doesn't rotate 1
Starter motor rotates but engine does not turn over 2
Starter works but engine won't turn over (seized) 3
No fuel flow 4
Engine flooded 5
No spark or weak spark 6
Compression low 7
Stalls after starting 8
Rough idle 9

Poor running at low speed
Spark weak 10
Fuel/air mixture incorrect 11
Compression low 12
Poor acceleration 13

Poor running or no power at high speed
Firing incorrect 14
Fuel/air mixture incorrect 15
Compression low 16
Knocking or pinging 17
Miscellaneous causes 18

Overheating
Engine overheats 19
Firing incorrect 20
Fuel/air mixture incorrect 21
Compression too high 22
Engine load excessive 23
Lubrication inadequate 24
Miscellaneous causes 25

Clutch problems
Clutch slipping 26
Clutch not disengaging completely 27

Gear shifting problems
Doesn't go into gear, or lever doesn't return 28
Jumps out of gear 29
Overshifts 30

Abnormal engine noise
Knocking or pinging 31
Piston slap or rattling 32
Valve noise 33
Other noise 34

Abnormal driveline noise
Clutch noise 35
Transmission noise 36
Final drive noise 37

Abnormal frame and suspension noise
Front end noise 38
Shock absorber noise 39
Disc brake noise 40

Oil level indicator light comes on
Engine lubrication system 41
Electrical system 42

Excessive exhaust smoke
White smoke 43
Black smoke 44
Brown smoke 45

Poor handling or stability
Handlebar hard to turn 46
Handlebar shakes or vibrates excessively 47
Handlebar pulls to one side 48
Poor shock absorbing qualities 49

Braking problems
Brakes are spongy, don't hold 50
Brake lever pulsates 51
Brakes drag 52

Electrical problems
Battery dead or weak 53
Battery overcharged 54

Engine doesn't start or is difficult to start

1 Starter motor does not rotate

1 Engine kill switch Off.
2 Fuse blown. Check fuse (Chapter 8).
3 Battery voltage low. Check and recharge battery (Chapter 8).
4 Starter motor defective. Make sure the wiring to the starter is secure. Test starter relay (Chapter 8). If the relay is good, then the fault is in the wiring or motor.
5 Starter relay faulty. Check it according to the procedure in Chapter 8.
6 Starter switch not contacting. The contacts could be wet, corroded or dirty. Disassemble and clean the switch (Chapter 8).
7 Wiring open or shorted. Check all wiring connections and harnesses to make sure that they are dry, tight and not corroded. Also check for broken or frayed wires that can cause a short to ground (see wiring diagram, Chapter 8).
8 Ignition (main) switch defective. Check the switch according to the procedure in Chapter 8. Replace the switch with a new one if it is defective.
9 Engine kill switch defective. Check for wet, dirty or corroded contacts. Clean or replace the switch as necessary (Chapter 8).

2 Starter motor rotates but engine does not turn over

1 Starter motor clutch defective. Inspect and repair or replace (Chapter 8).
2 Damaged starter reduction gears. Inspect and replace the damaged parts (Chapter 8).

3 Starter works but engine won't turn over (seized)

Seized engine caused by one or more internally damaged components. Failure due to wear, abuse or lack of lubrication. Damage can include seized valves, valve lifters, camshaft, piston, crankshaft, connecting rod bearings, or transmission gears or bearings. Refer to Chapter 2 for engine disassembly.

4 No fuel flow

1 No fuel in tank.
2 Tank cap air vent obstructed. Usually caused by dirt or water. Remove it and clean the cap vent hole.
3 Clogged strainer in fuel tap or inside fuel tank. Remove and clean the strainer(s) (Chapter 1).
4 Fuel line clogged. Pull the fuel line loose and carefully blow through it.
5 Inlet needle valve clogged. A very bad batch of fuel with an unusual additive may have been used, or some other foreign material has entered the tank. Many times after a machine has been stored for many months without running, the fuel turns to a varnish-like liquid and forms deposits on the inlet needle valve and jets. The carburetor should be removed and overhauled if draining the float chamber doesn't solve the problem.

5 Engine flooded

1 Float level too high. Check as described in Chapter 3 and replace the float if necessary.
2 Inlet needle valve worn or stuck open. A piece of dirt, rust or other debris can cause the inlet needle to seat improperly, causing excess fuel to be admitted to the float bowl. In this case, the float chamber should be cleaned and the needle and seat inspected. If the needle and seat are worn, then the leaking will persist and the parts should be replaced with new ones (Chapter 3).
3 Starting technique incorrect. Under normal circumstances (i.e., if all the carburetor functions are sound) the machine should start with little or no throttle. When the engine is cold, the choke should be operated and the engine started without opening the throttle. When the engine is at operating temperature, only a very slight amount of throttle should be necessary. If the engine is flooded, turn the fuel tap off and hold the throttle open while cranking the engine. This will allow additional air to reach the cylinder. Remember to turn the fuel tap back on after the engine starts.

6 No spark or weak spark

1 Ignition switch Off.
2 Engine kill switch turned to the Off position.
3 Battery voltage low. Check and recharge battery as necessary (Chapter 8).
4 Spark plug dirty, defective or worn out. Locate reason for fouled plug using spark plug condition chart and follow the plug maintenance procedures in Chapter 1.
5 Spark plug cap or secondary (HT) wiring faulty. Check condition. Replace either or both components if cracks or deterioration are evident (Chapter 4).
6 Spark plug cap not making good contact. Make sure that the plug cap fits snugly over the plug end.
7 Ignition control module defective. Check the unit, referring to Chapter 4 for details.
8 Ignition signal generator defective. Check the unit, referring to Chapter 4 for details.
9 Ignition coil defective. Check the coil, referring to Chapter 4.
10 Ignition or kill switch shorted. This is usually caused by water, corrosion, damage or excessive wear. The kill switch can be disassembled and cleaned with electrical contact cleaner. If cleaning does not help, replace the switches (Chapter 8).
11 Wiring shorted or broken between:
   a) *Ignition switch and engine kill switch (or blown fuse)*
   b) *Ignition control unit and engine kill switch*
   c) *Ignition control unit and ignition coil*
   d) *Ignition coil and plug*
   e) *Ignition control unit and ignition signal generator*

   Make sure that all wiring connections are clean, dry and tight. Look for chafed and broken wires (Chapters 4 and 8).

7 Compression low

1 Spark plug loose. Remove the plug and inspect the threads. Reinstall and tighten to the specified torque (Chapter 1).
2 Cylinder head not sufficiently tightened down. If the cylinder head is suspected of being loose, then there's a chance that the gasket or head is damaged if the problem has persisted for any length of time. The head nuts and bolts should be tightened to the proper torque in the correct sequence (Chapter 2).
3 Improper valve clearance. This means that the valve is not closing completely and compression pressure is leaking past the valve. Check and adjust the valve clearances (Chapter 1).
4 Cylinder and/or piston worn. Excessive wear will cause compression pressure to leak past the rings. This is usually accompanied by worn rings as well. A top end overhaul is necessary (Chapter 2).
5 Piston rings worn, weak, broken, or sticking. Broken or sticking piston rings usually indicate a lubrication or carburetion problem that causes excess carbon deposits or seizures to form on the pistons and rings. Top end overhaul is necessary (Chapter 2).
6 Piston ring-to-groove clearance excessive. This is caused by

excessive wear of the piston ring lands. Piston replacement is necessary (Chapter 2).

7   Cylinder head gasket damaged. If the head is allowed to become loose, or if excessive carbon build-up on a piston crown and combustion chamber causes extremely high compression, the head gasket may leak. Retorquing the head is not always sufficient to restore the seal, so gasket replacement is necessary (Chapter 2).

8   Cylinder head warped. This is caused by overheating or improperly tightened head nuts and bolts. Machine shop resurfacing or head replacement is necessary (Chapter 2).

9   Valve spring broken or weak. Caused by component failure or wear; the spring(s) must be replaced (Chapter 2).

10   Valve not seating properly. This is caused by a bent valve (from over-revving or improper valve adjustment), burned valve or seat (improper carburetion) or an accumulation of carbon deposits on the seat (from carburetion or lubrication problems). The valves must be cleaned and/or replaced and the seats serviced if possible (Chapter 2).

8   Stalls after starting

1   Improper choke action. Make sure the choke cable is getting a full stroke and staying in the out position.

2   Ignition malfunction. See Chapter 4.

3   Carburetor malfunction. See Chapter 3.

4   Fuel contaminated. The fuel can be contaminated with either dirt or water, or can change chemically if the machine is allowed to sit for several months or more. Drain the tank and float bowl and refill with fresh fuel (Chapter 3).

5   Intake air leak. Check for loose carburetor-to-intake joint connections or loose carburetor top (Chapter 3).

6   Engine idle speed incorrect. Turn throttle stop screw until the engine idles at the specified rpm (Chapter 1).

9   Rough idle

1   Ignition malfunction. See Chapter 4.

2   Idle speed incorrect. See Chapter 1.

3   Carburetor malfunction. See Chapter 3.

4   Idle fuel/air mixture incorrect. See Chapter 3.

5   Fuel contaminated. The fuel can be contaminated with either dirt or water, or can change chemically if the machine is allowed to sit for several months or more. Drain the tank and float bowls (Chapter 3).

6   Intake air leak. Check for loose carburetor-to-intake joint connections, loose or missing vacuum gauge access port cap or hose, or loose carburetor top (Chapter 3).

7   Air cleaner clogged. Service or replace air cleaner element (Chapter 1).

Poor running at low speed

10   Spark weak

1   Battery voltage low. Check and recharge battery (Chapter 8).

2   Spark plug fouled, defective or worn out. Refer to Chapter 1 for spark plug maintenance.

3   Spark plug cap or secondary (HT) wiring defective. Refer to Chapters 1 and 4 for details on the ignition system.

4   Spark plug cap not making contact.

5   Incorrect spark plug. Wrong type, heat range or cap configuration. Check and install correct plug listed in Chapter 1. A cold plug or one with a recessed firing electrode will not operate at low speeds without fouling.

6   Ignition control module defective. See Chapter 4.

7   Pulse generator defective. See Chapter 4.

8   Ignition coil defective. See Chapter 4.

11   Fuel/air mixture incorrect

1   Pilot screw out of adjustment (Chapter 3).

2   Pilot jet or air passage clogged. Remove and overhaul the carburetor (Chapter 3).

3   Air bleed holes clogged. Remove carburetor and blow out all passages (Chapter 3).

4   Air cleaner clogged, poorly sealed or missing.

5   Air cleaner-to-carburetor boot poorly sealed. Look for cracks, holes or loose clamps and replace or repair defective parts.

6   Float level too high or too low. Check and replace the float if necessary (Chapter 3).

7   Fuel tank air vent obstructed. Make sure that the air vent passage in the filler cap is open.

8   Carburetor intake joint loose. Check for cracks, breaks, tears or loose clamps or bolts. Repair or replace the rubber boot and its O-ring.

12   Compression low

1   Spark plug loose. Remove the plug and inspect the threads. Reinstall and tighten to the specified torque (Chapter 1).

2   Cylinder head not sufficiently tightened down. If the cylinder head is suspected of being loose, then there's a chance that the gasket and head are damaged if the problem has persisted for any length of time. The head nuts and bolts should be tightened to the proper torque in the correct sequence (Chapter 2).

3   Improper valve clearance. This means that the valve is not closing completely and compression pressure is leaking past the valve. Check and adjust the valve clearances (Chapter 1).

4   Cylinder and/or piston worn. Excessive wear will cause compression pressure to leak past the rings. This is usually accompanied by worn rings as well. A top end overhaul is necessary (Chapter 2).

5   Piston rings worn, weak, broken, or sticking. Broken or sticking piston rings usually indicate a lubrication or carburetion problem that causes excess carbon deposits or seizures to form on the pistons and rings. Top end overhaul is necessary (Chapter 2).

6   Piston ring-to-groove clearance excessive. This is caused by excessive wear of the piston ring lands. Piston replacement is necessary (Chapter 2).

7   Cylinder head gasket damaged. If the head is allowed to become loose, or if excessive carbon build-up on the piston crown and combustion chamber causes extremely high compression, the head gasket may leak. Retorquing the head is not always sufficient to restore the seal, so gasket replacement is necessary (Chapter 2).

8   Cylinder head warped. This is caused by overheating or improperly tightened head nuts and bolts. Machine shop resurfacing or head replacement is necessary (Chapter 2).

9   Valve spring broken or weak. Caused by component failure or wear; the spring(s) must be replaced (Chapter 2).

10   Valve not seating properly. This is caused by a bent valve (from over-revving or improper valve adjustment), burned valve or seat (improper carburetion) or an accumulation of carbon deposits on the seat (from carburetion, lubrication problems). The valves must be cleaned and/or replaced and the seats serviced if possible (Chapter 2).

13   Poor acceleration

1   Carburetor leaking or dirty. Overhaul the carburetor (Chapter 3).

2   Timing not advancing. The pulse generator or the ignition control module may be defective. If so, they must be replaced with new ones, as they can't be repaired.

3   Engine oil viscosity too high. Using a heavier oil than that recommended in Chapter 1 can damage the oil pump or lubrication

system and cause drag on the engine.

4   Brakes dragging. Usually caused by debris which has entered the brake piston sealing boots (front brakes), corroded wheel cylinders (drum brakes) or calipers (disc brakes) or from a warped drum or bent axle. Repair as necessary (Chapter 6).

Poor running or no power at high speed

14   Firing incorrect

1   Air cleaner restricted. Clean or replace element (Chapter 1).
2   Spark plug fouled, defective or worn out. See Chapter 1 for spark plug maintenance.
3   Spark plug cap or secondary (HT) wiring defective. See Chapters 1 and 4 for details of the ignition system.
4   Spark plug cap not in good contact. See Chapter 4.
5   Incorrect spark plug. Wrong type, heat range or cap configuration. Check and install correct plugs listed in Chapter 1. A cold plug or one with a recessed firing electrode will not operate at low speeds without fouling.
6   Ignition control module defective. See Chapter 4.
7   Ignition coil defective. See Chapter 4.

15   Fuel/air mixture incorrect

1   Pilot screw out of adjustment. See Chapter 3 for adjustment procedures.
2   Main jet clogged. Dirt, water or other contaminants can clog the main jets. Clean the fuel tap strainer and in-tank strainer, the float bowl area, and the jets and carburetor orifices (Chapter 3).
3   Main jet wrong size. The standard jetting is for sea level atmospheric pressure and oxygen content. See Chapter 3 for high altitude adjustments.
4   Throttle shaft-to-carburetor body clearance excessive. Refer to Chapter 3 for inspection and part replacement procedures.
5   Air bleed holes clogged. Remove and overhaul carburetor (Chapter 3).
6   Air cleaner clogged, poorly sealed, or missing.
7   Air cleaner-to-carburetor boot poorly sealed. Look for cracks, holes or loose clamps, and replace or repair defective parts.
8   Float level too high or too low. Check float level and replace the float if necessary (Chapter 3).
9   Fuel tank air vent obstructed. Make sure the air vent passage in the filler cap is open.
10   Carburetor intake joint loose. Check for cracks, breaks, tears or loose clamps or bolts. Repair or replace the rubber boots (Chapter 3).
11   Fuel tap clogged. Remove the tap and clean it (Chapter 1).
12   Fuel line clogged. Pull the fuel line loose and carefully blow through it.

16   Compression low

1   Spark plug loose. Remove the plug and inspect the threads. Reinstall and tighten to the specified torque (Chapter 1).
2   Cylinder head not sufficiently tightened down. If the cylinder head is suspected of being loose, then there's a chance that the gasket and head are damaged if the problem has persisted for any length of time. The head nuts and bolts should be tightened to the proper torque in the correct sequence (Chapter 2).
3   Improper valve clearance. This means that the valve is not closing completely and compression pressure is leaking past the valve. Check and adjust the valve clearances (Chapter 1).
4   Cylinder and/or piston worn. Excessive wear will cause compression pressure to leak past the rings. This is usually accompanied by worn rings as well. A top end overhaul is necessary (Chapter 2).

5   Piston rings worn, weak, broken, or sticking. Broken or sticking piston rings usually indicate a lubrication or carburetion problem that causes excess carbon deposits or seizures to form on the pistons and rings. Top end overhaul is necessary (Chapter 2).
6   Piston ring-to-groove clearance excessive. This is caused by excessive wear of the piston ring lands. Piston replacement is necessary (Chapter 2).
7   Cylinder head gasket damaged. If a head is allowed to become loose, or if excessive carbon build-up on the piston crown and combustion chamber causes extremely high compression, the head gasket may leak. Retorquing the head is not always sufficient to restore the seal, so gasket replacement is necessary (Chapter 2).
8   Cylinder head warped. This is caused by overheating or improperly tightened head nuts and bolts. Machine shop resurfacing or head replacement is necessary (Chapter 2).
9   Valve spring broken or weak. Caused by component failure or wear; the spring(s) must be replaced (Chapter 2).
10   Valve not seating properly. This is caused by a bent valve (from over-revving or improper valve adjustment), burned valve or seat (improper carburetion) or an accumulation of carbon deposits on the seat (from carburetion or lubrication problems). The valves must be cleaned and/or replaced and the seats serviced if possible (Chapter 2).

17   Knocking or pinging

1   Carbon build-up in combustion chamber. Use of a fuel additive that will dissolve the adhesive bonding the carbon particles to the crown and chamber is the easiest way to remove the build-up. Otherwise, the cylinder head will have to be removed and decarbonized (Chapter 2).
2   Incorrect or poor quality fuel. Old or improper grades of fuel can cause detonation. This causes the piston to rattle, thus the knocking or pinging sound. Drain old fuel and always use the recommended fuel grade.
3   Spark plug heat range incorrect. Uncontrolled detonation indicates the plug heat range is too hot. The plug in effect becomes a glow plug, raising cylinder temperatures. Install the proper heat range plug (Chapter 1).
4   Improper air/fuel mixture. This will cause the cylinder to run hot, which leads to detonation. Clogged jets or an air leak can cause this imbalance. See Chapter 3.

18   Miscellaneous causes

1   Throttle valve doesn't open fully. Adjust the cable slack (Chapter 1).
2   Clutch slipping. May be caused by improperly adjustment or loose or worn clutch components. Refer to Chapter 1 for adjustment or Chapter 2 for cable replacement and clutch overhaul procedures.
3   Timing not advancing.
4   Engine oil viscosity too high. Using a heavier oil than the one recommended in Chapter 1 can damage the oil pump or lubrication system and cause drag on the engine.
5   Brakes dragging. Usually caused by debris which has entered the brake piston sealing boot, or from a warped disc or bent axle. Repair as necessary.

Overheating

19   Engine overheats

1   Engine oil level low. Check and add oil (Chapter 1).
2   Wrong type of oil. If you're not sure what type of oil is in the engine, drain it and fill with the correct type (Chapter 1).

3   Air leak at carburetor intake joints. Check and tighten or replace as necessary (Chapter 3).

4   Fuel level low. Check and adjust if necessary (Chapter 3).

5   Worn oil pump or clogged oil passages. Replace pump or clean passages as necessary.

6   Clogged external oil line. Remove and check for foreign material (see Chapter 2).

7   Carbon build-up in combustion chambers. Use of a fuel additive that will dissolve the adhesive bonding the carbon particles to the piston crown and chambers is the easiest way to remove the build-up. Otherwise, the cylinder head will have to be removed and decarbonized (Chapter 2).

8   Operation in high ambient temperatures.

9   Electric shift system problems (models so equipped). Check the system (Chapter 8).

20   Firing incorrect

1   Spark plug fouled, defective or worn out. See Chapter 1 for spark plug maintenance.

2   Incorrect spark plug (see Chapter 1).

3   Faulty ignition coil(s) (Chapter 4).

21   Fuel/air mixture incorrect

1   Pilot screw out of adjustment (Chapter 3).

2   Main jet clogged. Dirt, water and other contaminants can clog the main jets. Clean the fuel tap strainer and in-tank strainer, the float bowl area and the jets and carburetor orifices (Chapter 3).

3   Main jet wrong size. The standard jetting is for sea level atmospheric pressure and oxygen content. See Chapter 3 for high altitude settings.

4   Air cleaner poorly sealed or missing.

5   Air cleaner-to-carburetor boot poorly sealed. Look for cracks, holes or loose clamps and replace or repair.

6   Fuel level too low. Check float level and replace the float if necessary (Chapter 3).

7   Fuel tank air vent obstructed. Make sure that the air vent passage in the filler cap is open.

8   Carburetor intake joint loose. Check for cracks, breaks, tears or loose clamps or bolts. Repair or replace the rubber boot and its O-ring (Chapter 3).

22   Compression too high

1   Carbon build-up in combustion chamber. Use of a fuel additive that will dissolve the adhesive bonding the carbon particles to the piston crown and chamber is the easiest way to remove the build-up. Otherwise, the cylinder head will have to be removed and decarbonized (Chapter 2).

2   Improperly machined head surface or installation of incorrect gasket during engine assembly.

23   Engine load excessive

1   Clutch slipping. Can be caused by damaged, loose or worn clutch components. Refer to Chapter 2 for overhaul procedures.

2   Engine oil level too high. The addition of too much oil will cause pressurization of the crankcase and inefficient engine operation. Check Specifications and drain to proper level (Chapter 1).

3   Engine oil viscosity too high. Using a heavier oil than the one recommended in Chapter 1 can damage the oil pump or lubrication system as well as cause drag on the engine.

4   Brakes dragging. Usually caused by debris which has entered the wheel cylinders or calipers, corroded wheel cylinders or calipers, or from a warped drum or disc or bent axle. Repair as necessary (Chapter 6).

24   Lubrication inadequate

1   Engine oil level too low. Friction caused by intermittent lack of lubrication or from oil that is overworked can cause overheating. The oil provides a definite cooling function in the engine. Check the oil level (Chapter 1).

2   Poor quality engine oil or incorrect viscosity or type. Oil is rated not only according to viscosity but also according to type. Some oils are not rated high enough for use in this engine. Check the Specifications section and change to the correct oil (Chapter 1).

3   Camshaft or journals worn. Excessive wear causing drop in oil pressure. Replace cam or cylinder head. Abnormal wear could be caused by oil starvation at high rpm from low oil level or improper viscosity or type of oil (Chapter 1).

4   Crankshaft and/or bearings worn. Same problems as paragraph 3. Check and replace crankshaft assembly if necessary (Chapter 2).

25   Miscellaneous causes

Modification to exhaust system. Most aftermarket exhaust systems cause the engine to run leaner, which makes it run hotter. When installing an accessory exhaust system, always rejet the carburetor.

Clutch problems

26   Clutch slipping

1   Change clutch friction plates worn or warped. Overhaul the change clutch assembly (Chapter 2).

2   Change clutch steel plates worn or warped (Chapter 2).

3   Change clutch spring(s) broken or weak. Old or heat-damaged spring(s) (from slipping clutch) should be replaced with new ones (Chapter 2).

4   Change clutch release mechanism defective. Replace any defective parts (Chapter 2).

5   Change clutch center or housing unevenly worn. This causes improper engagement of the plates. Replace the damaged or worn parts (Chapter 2).

6   Centrifugal clutch weight linings or drum worn (Chapter 2).

27   Clutch not disengaging completely

1   Change clutch improperly adjusted (see Chapter 1).

2   Change clutch plates warped or damaged. This will cause clutch drag, which in turn will cause the machine to creep. Overhaul the clutch assembly (Chapter 2).

3   Sagged or broken change clutch spring(s). Check and replace the spring(s) (Chapter 2).

4   Engine oil deteriorated. Old, thin, worn out oil will not provide proper lubrication for the discs, causing the change clutch to drag. Replace the oil and filter (Chapter 1).

5   Engine oil viscosity too high. Using a thicker oil than recommended in Chapter 1 can cause the change clutch plates to stick together, putting a drag on the engine. Change to the correct viscosity oil (Chapter 1).

6   Change clutch housing seized on shaft. Lack of lubrication, severe wear or damage can cause the housing to seize on the shaft. Overhaul of the clutch, and perhaps transmission, may be necessary to repair the damage (Chapter 2).

7   Change clutch release mechanism defective. Worn or damaged release mechanism parts can stick and fail to apply force to the pressure plate. Overhaul the release mechanism (Chapter 2).
8   Loose change clutch center nut. Causes housing and center misalignment putting a drag on the engine. Engagement adjustment continually varies. Overhaul the clutch assembly (Chapter 2).
9   Weak or broken centrifugal clutch springs (Chapter 2).

Gear shifting problems

28   Doesn't go into gear or lever doesn't return

1   Clutch not disengaging. See Section 27.
2   Shift fork(s) bent or seized. May be caused by lack of lubrication. Overhaul the transmission (Chapter 2).
3   Gear(s) stuck on shaft. Most often caused by a lack of lubrication or excessive wear in transmission bearings and bushings. Overhaul the transmission (Chapter 2).
4   Shift drum binding. Caused by lubrication failure or excessive wear. Replace the drum and bearing (Chapter 2).
5   Shift lever return spring weak or broken (Chapter 2).
6   Shift lever broken. Splines stripped out of lever or shaft, caused by allowing the lever to get loose. Replace necessary parts (Chapter 2).
7   Shift mechanism pawl broken or worn. Full engagement and rotary movement of shift drum results. Replace shaft assembly (Chapter 2).
8   Pawl spring broken. Allows pawl to float, causing sporadic shift operation. Replace spring (Chapter 2).

29   Jumps out of gear

1   Shift fork(s) worn. Overhaul the transmission (Chapter 2).
2   Gear groove(s) worn. Overhaul the transmission (Chapter 2).
3   Gear dogs or dog slots worn or damaged. The gears should be inspected and replaced. No attempt should be made to service the worn parts.

30   Overshifts

1   Pawl spring weak or broken (Chapter 2).
2   Shift drum stopper lever not functioning (Chapter 2).

Abnormal engine noise

31   Knocking or pinging

1   Carbon build-up in combustion chamber. Use of a fuel additive that will dissolve the adhesive bonding the carbon particles to the piston crown and chamber is the easiest way to remove the build-up. Otherwise, the cylinder head will have to be removed and decarbonized (Chapter 2).
2   Incorrect or poor quality fuel. Old or improper fuel can cause detonation. This causes the pistons to rattle, thus the knocking or pinging sound. Drain the old fuel (Chapter 3) and always use the recommended grade fuel (Chapter 1).
3   Spark plug heat range incorrect. Uncontrolled detonation indicates that the plug heat range is too hot. The plug in effect becomes a glow plug, raising cylinder temperatures. Install the proper heat range plug (Chapter 1).
4   Improper air/fuel mixture. This will cause the cylinder to run hot and lead to detonation. Clogged jets or an air leak can cause this imbalance. See Chapter 3.

32   Piston slap or rattling

1   Cylinder-to-piston clearance excessive. Caused by improper assembly. Inspect and overhaul top end parts (Chapter 2).
2   Connecting rod bent. Caused by over-revving, trying to start a badly flooded engine or from ingesting a foreign object into the combustion chamber. Replace the damaged parts (Chapter 2).
3   Piston pin or piston pin bore worn or seized from wear or lack of lubrication. Replace damaged parts (Chapter 2).
4   Piston ring(s) worn, broken or sticking. Overhaul the top end (Chapter 2).
5   Piston seizure damage. Usually from lack of lubrication or overheating. Replace the pistons and bore the cylinder, as necessary (Chapter 2).
6   Connecting rod upper or lower end clearance excessive. Caused by excessive wear or lack of lubrication. Replace worn parts.

33   Valve noise

1   Incorrect valve clearances. Adjust the clearances by referring to Chapter 1.
2   Valve spring broken or weak. Check and replace weak valve springs (Chapter 2).
3   Camshaft or cylinder head worn or damaged. Lack of lubrication at high rpm is usually the cause of damage. Insufficient oil or failure to change the oil at the recommended intervals are the chief causes.

34   Other noise

1   Cylinder head gasket leaking.
2   Exhaust pipe leaking at cylinder head connection. Caused by improper fit of pipe, damaged gasket or loose exhaust flange. All exhaust fasteners should be tightened evenly and carefully. Failure to do this will lead to a leak.
3   Crankshaft runout excessive. Caused by a bent crankshaft (from over-revving) or damage from an upper cylinder component failure.
4   Engine mounting bolts or nuts loose. Tighten all engine mounting bolts and nuts to the specified torque (Chapter 2).
5   Crankshaft bearings worn (Chapter 2).
6   Camshaft chain tensioner defective. Replace according to the procedure in Chapter 2.
7   Camshaft chain, sprockets or guides worn (Chapter 2).

Abnormal driveline noise

35   Clutch noise

1   Change clutch housing/friction plate clearance excessive (Chap-ter 2).
2   Loose or damaged change clutch pressure plate and/or bolts (Chapter 2).
3   Broken centrifugal clutch springs (Chapter 2).

36   Transmission noise

1   Bearings worn. Also includes the possibility that the shafts are worn. Overhaul the transmission (Chapter 2).
2   Gears worn or chipped (Chapter 2).
3   Metal chips jammed in gear teeth. Probably pieces from a broken clutch, gear or shift mechanism that were picked up by the gears. This

will cause early bearing failure (Chapter 2).
4   Engine oil level too low. Causes a howl from transmission. Also affects engine power and clutch operation (Chapter 1).

37 Differential noise

1   Differential oil level low (Chapter 1).
2   Differential gear lash out of adjustment. Checking and adjustment require special tools and skills and should be done by a Honda dealer.
3   Differential gears damaged or worn. Overhaul requires special tools and skills and should be done by a Honda dealer.

Abnormal chassis noise

38 Suspension noise

1   Spring weak or broken. Makes a clicking or scraping sound.
2   Steering shaft bearings worn or damaged. Clicks when braking. Check and replace as necessary (Chapter 5).
3   Shock absorber fluid level incorrect. Indicates a leak caused by defective seal. Shock will be covered with oil. Replace shock (Chapter 5).
4   Defective shock absorber with internal damage. This is in the body of the shock and can't be remedied. The shock must be replaced with a new one (Chapter 5).
5   Bent or damaged shock body. Replace the shock with a new one (Chapter 5).

39 Driveaxle noise

1   Worn or damaged outer joint. Makes clicking noise in turns. Check for cut or damaged seals and repair as necessary (see Chapter 5).
2   Worn or damaged inner joint. Makes knock or clunk when accelerating after coasting. Check for cut or damaged seals and repair as necessary (see Chapter 5).

40 Brake noise

1   Brake linings worn or contaminated. Can cause scraping or squealing. Replace the shoes or pads (Chapter 6).
2   Brake linings warped or worn unevenly. Can cause chattering. Replace the linings (Chapter 6).
3   Brake drum or disc out of round. Can cause chattering. Replace (Chapter 6).
4   Loose or worn knuckle or rear axle bearings. Check and replace as needed (Chapter 5).

Oil temperature indicator light comes on

41 Engine lubrication system

1   High oil temperature due to operation in high ambient temperatures. Shut the engine off and let it cool. Check the oil cooler for clogged fins or tubes and clean it as necessary.

42 Electrical system

1   Oil temperature sensor defective. Check the sensor according to the procedure in Chapter 8. Replace it if it's defective.

2   Oil temperature indicator light circuit defective. Check for pinched, shorted, disconnected or damaged wiring (Chapter 8).
3   Oil cooler fan not working correctly (see Chapter 8).

Excessive exhaust smoke

43 White smoke

1   Piston oil ring worn. The ring may be broken or damaged, causing oil from the crankcase to be pulled past the piston into the combustion chamber. Replace the rings with new ones (Chapter 2).
2   Cylinders worn, cracked, or scored. Caused by overheating or oil starvation. If worn or scored, the cylinders will have to be rebored and new pistons installed. If cracked, the cylinder block will have to be replaced (see Chapter 2).
3   Valve oil seal damaged or worn. Replace oil seals with new ones (Chapter 2).
4   Valve guide worn. Perform a complete valve job (Chapter 2).
5   Engine oil level too high, which causes the oil to be forced past the rings. Drain oil to the proper level (Chapter 1).
6   Head gasket broken between oil return and cylinder. Causes oil to be pulled into the combustion chamber. Replace the head gasket and check the head for warpage (Chapter 2).
7   Abnormal crankcase pressurization, which forces oil past the rings. Clogged breather or hoses usually the cause (Chapter 2).

44 Black smoke

1   Air cleaner clogged. Clean or replace the element (Chapter 1).
2   Main jet too large or loose. Compare the jet size to the Specifications (Chapter 3).
3   Choke stuck, causing fuel to be pulled through choke circuit (Chapter 3).
4   Fuel level too high. Check the float level and replace the float if necessary (Chapter 3).
5   Inlet needle held off needle seat. Clean the float chamber and fuel line and replace the needle and seat if necessary (Chapter 3).

45 Brown smoke

1   Main jet too small or clogged. Lean condition caused by wrong size main jet or by a restricted orifice. Clean float chamber and jets and compare jet size to Specifications (Chapter 3).
2   Fuel flow insufficient. Fuel inlet needle valve stuck closed due to chemical reaction with old fuel. Float level incorrect; check and replace float if necessary. Restricted fuel line. Clean line and float chamber.
3   Carburetor intake tube loose (Chapter 3).
4   Air cleaner poorly sealed or not installed (Chapter 1).

Poor handling or stability

46 Handlebar hard to turn

1   Steering shaft nut too tight (Chapter 5).
2   Lower bearing or upper bushing damaged. Roughness can be felt as the bars are turned from side-to-side. Replace bearing and bushing (Chapter 5).
3   Steering shaft bearing lubrication inadequate. Caused by grease getting hard from age or being washed out by high pressure car washes. Remove steering shaft and replace bearing (Chapter 5).
4   Steering shaft bent. Caused by a collision, hitting a pothole or by rolling the machine. Replace damaged part. Don't try to straighten the steering shaft (Chapter 5).
5   Front tire air pressure too low (Chapter 1).

47 Handlebar shakes or vibrates excessively

1   Tires worn or out of balance (Chapter 1 or 6).
2   Swingarm bearings worn. Replace worn bearings by referring to Chapter 6.
3   Wheel rim(s) warped or damaged. Inspect wheels (Chapter 6).
4   Wheel bearings worn. Worn front or rear wheel bearings can cause poor tracking. Worn front bearings will cause wobble (Chapter 6).
5   Wheel hubs installed incorrectly (Chapter 5 or Chapter 6).
6   Handlebar clamp bolts or bracket nuts loose (Chapter 5).
7   Steering shaft nut or bolts loose. Tighten them to the specified torque (Chapter 5).
8   Motor mount bolts loose. Will cause excessive vibration with increased engine rpm (Chapter 2).

48 Handlebar pulls to one side

1   Uneven tire pressures (Chapter 1).
2   Frame bent. Definitely suspect this if the machine has been rolled. May or may not be accompanied by cracking near the bend. Replace the frame (Chapter 5).
3   Wheel out of alignment. Caused by incorrect toe-in adjustment (Chapter 1) or bent tie-rod (Chapter 5).
4   Swingarm bent or twisted. Caused by age (metal fatigue) or impact damage. Replace the swingarm (Chapter 5).
5   Steering shaft bent. Caused by impact damage or by rolling the vehicle. Replace the steering stem (Chapter 5).

49 Poor shock absorbing qualities

1   Too hard:
a) Shock internal damage.
b) Tire pressure too high (Chapters 1 and 6).
2   Too soft:
a) Shock oil insufficient and/or leaking (Chapter 5).
d) Fork springs weak or broken (Chapter 5).

Braking problems

50 Front brakes are spongy, don't hold

1   Air in brake line. Caused by inattention to master cylinder fluid level or by leakage. Locate problem and bleed brakes (Chapter 6).
2   Linings worn (Chapters 1 and 6).
3   Brake fluid leak. See paragraph 1.
4   Contaminated linings or pads. Caused by contamination with oil, grease, brake fluid, etc. Clean or replace linings. Clean drum or disc thoroughly with brake cleaner (Chapter 6).
5   Brake fluid deteriorated. Fluid is old or contaminated. Drain system, replenish with new fluid and bleed the system (Chapter 6).

6   Master cylinder internal parts worn or damaged causing fluid to bypass (Chapter 6).
7   Master cylinder bore scratched by foreign material or broken spring. Repair or replace master cylinder (Chapter 6).
8   Drum or disc warped. Replace (Chapter 6).

51 Brake lever or pedal pulsates

1   Axle bent. Replace axle (Chapter 5).
2   Wheel warped or otherwise damaged (Chapter 6).
3   Hub or axle bearings damaged or worn (Chapter 6).
4   Brake drum or disc out of round. Replace (Chapter 6).

52 Brakes drag

1   Master cylinder piston seized. Caused by wear or damage to piston or cylinder bore (Chapter 6).
2   Lever balky or stuck. Check pivot and lubricate (Chapter 6).
3   Wheel cylinder or caliper piston seized in bore. Caused by wear or ingestion of dirt past deteriorated seal (Chapter 6).
4   Brake shoes or pads damaged. Lining material separated from shoes. Usually caused by faulty manufacturing process or from contact with chemicals. Replace shoes (Chapter 6).
5   Shoes improperly installed (Chapter 6).
6   Rear brake pedal or lever free play insufficient (Chapter 1).
7   Rear brake springs weak. Replace brake springs (Chapter 6).

Electrical problems

53 Battery dead or weak

1   Battery faulty. Caused by sulfated plates which are shorted through sedimentation or low electrolyte level. Also, broken battery terminal making only occasional contact (Chapter 8).
2   Battery cables making poor contact (Chapter 8).
3   Load excessive. Caused by addition of high wattage lights or other electrical accessories.
4   Ignition switch defective. Switch either grounds/earths internally or fails to shut off system. Replace the switch (Chapter 8).
5   Regulator/rectifier defective (Chapter 8).
6   Stator coil open or shorted (Chapter 8).
7   Wiring faulty. Wiring grounded or connections loose in ignition, charging or lighting circuits (Chapter 8).

54 Battery overcharged

1   Regulator/rectifier defective. Overcharging is noticed when battery gets excessively warm or boils over (Chapter 8).
2   Battery defective. Replace battery with a new one (Chapter 8).
3   Battery amperage too low, wrong type or size. Install manufacturer's specified amp-hour battery to handle charging load (Chapter 8).

Notes

Chapter 1
Tune-up and routine maintenance

Contents

	Section
Air cleaner - filter element and drain tube cleaning	16
Battery - check	4
Brake lever and pedal freeplay - check and adjustment	6
Brake system - general check	5
Choke - operation check	12
Clutch - check and freeplay adjustment	9
Differential oil - change	15
Driveaxle boots (4wd models) - check	10
Engine oil/filter - change	14
Exhaust system - inspection and spark arrester cleaning	18
Fasteners - check	22
Fluid levels - check	3

	Section
Fuel system - check and filter cleaning	17
Honda TRX250/350 Routine maintenance schedule	1
Idle speed - check and adjustment	21
Introduction to tune-up and routine maintenance	2
Lubrication - general	13
Reverse lock system - check and adjustment	8
Spark plug - replacement	19
Steering system - inspection and toe-in adjustment	24
Suspension - check	23
Throttle operation/grip freeplay - check and adjustment	11
Tires/wheels - general check	7
Valve clearances - check and adjustment	20

Specifications

Engine

Spark plugs
  Type
    Rancher models
      Standard .... NGK DPR7EA-9 or ND X22EPR-U9
      Extended high speed riding .... NGK DPR6EA-9 or ND X20EPR-U9
    Recon models
      Standard .... NGK DPR8EA-9 or ND X24EPR-U9
      Cold weather (below 5-degrees C/40-degrees F) .... NGK DPR7EA-9 or ND X22EPR-U9
      Extended high speed riding .... NGK DPR9EA-9 or ND X27EPR-U9
    250EX models
      Standard .... NGK DPR8EA-9 or ND X24EPR-U9
      Cold weather (below 5-degrees C/40-degrees F) .... NGK DPR7EA-9 or ND X22EPR-U9
  Gap (all models) .... 0.8 to 0.0 mm (0.031 to 0.005 inch)
Engine idle speed .... 1400 +/- 100 rpm
Valve clearance (COLD engine, intake and exhaust)
  Rancher models .... 0.15 mm (0.006 inch)
  Recon and 250EX models .... 0.13 mm (0.005 inch)

Miscellaneous

Front brake shoe lining thickness (Rancher and Recon models)	
New	4 mm (0.16 inch)
Wear limit	2 mm (0.08 inch)
Front brake pad wear (250EX models)	See text
Rear brake shoe lining thickness	
New	
Rancher models	5 mm (0.20 inch)
Recon models	4.5 mm (0.18 inch)
Wear limit	Shown by indicator
Front brake lever freeplay	
Rancher and Recon models	25 to 30 mm (1 to 1-3/16 inch)
250EX models	10 to 25 mm (3/8 to 1 inch)
Rear brake lever freeplay	15 to 20 mm (5/8 to 3/4 inch)
Rear brake pedal freeplay	15 to 20 mm (5/8 to 3/4 inch)
Reverse selector lever freeplay	2 to 4 mm (1/16 to 1/8 inch)
Throttle lever freeplay	3 to 8 mm (1/8 to 5/16 inch)
Clutch lever freeplay (2006 and later TRX250EX)	10 to 20 mm (3/8 to 3/4 inch)
Choke freeplay	Not adjustable
Minimum tire tread depth	4 mm (0.16 inch)
Tire pressures (cold) (1)	
Rancher 2WD and 1997 through 2007 Recon models	
Minimum	2.5 psi
Standard	2.9 psi
Maximum	3.3 psi
Rear tires with cargo	2.9 psi
2008 and later Recon models	2.9 psi
Rancher 4WD models	
Minimum	3.2 psi
Standard	3.6 psi
Maximum	4.0 psi
Rear tires with cargo	3.6 psi
250EX models (front)	
Minimum	3.8 psi
Standard	4.4 psi
Maximum	5.0 psi
With cargo	4.4 psi
250EX models (rear)	
Minimum	2.5 psi
Standard	2.9 psi
Maximum	3.3 psi
With cargo	2.9 psi

Torque specifications

Engine oil drain plug	25 Nm (18 ft-lbs)
Oil filter cover bolts (Rancher models)	10 Nm (84 inch-lbs)
Clutch adjusting screw locknut	22 Nm (16 ft-lbs)
Timing hole cap	10 Nm (84 inch-lbs)
Valve adjusting screw locknuts	17 Nm (144 inch-lbs)
Spark plugs	
All Rancher, 2004 and earlier Recon/250EX	18 Nm (156 inch-lbs)
2005 and later Recon/250EX	22 Nm (16 ft-lbs)
Differential filler, drain and check plugs	12 Nm (108 inch-lbs)
Tie rod locknuts	54 Nm (40 ft-lbs)

Recommended lubricants and fluids

Engine/transmission oil	
Type	Honda GN4 or equivalent API grade SG or higher multigrade oil meeting JASO standard MA
Viscosity	
2005 and earlier	10W-40
2006 and later	10W-30
Capacity	
Rancher models	
Oil change only..	1.95 liters (2.06 US qt, 3.44 Imp pt)
With filter change..	2.0 liters (2.1 US qt, 3.6 Imp pt)
After engine overhaul..	2.5 liters (2.6 US qt, 4.4 Imp pt)
Recon models (1997 through 2001)	
Oil change (drain and fill)	1.6 liters (1.7 US qt, 2.8 Imp pt)
After overhaul	1.8 liters (1.9 US qt, 3.2 Imp pt)
Recon models (2002 and later)	
Oil change (drain and fill)	1.5 liters (1.6 US qt, 2.6 Imp pt)
After overhaul	1.9 liters (2.0 US qt, 3.4 Imp pt)
250EX models	
Oil change (drain and fill)	1.6 liters (1.7 US qt, 2.8 Imp pt)
After overhaul	1.9 liters (2.0 US qt, 3.4 Imp pt)
Differential oil	
Type	Hypoid gear oil
Viscosity..	SAE 80
Front differential capacity..	241 cc (8.2 US fl oz, 8.5 Imp oz)
Rear differential capacity	
Rancher models..	85 cc (2.9 US fl oz, 3.0 Imp oz)
Recon models...	80 cc (2.7 US fl oz, 2.8 Imp oz)
250EX models...	75 cc (2.5 US fl oz, 2.6 Imp oz)
Brake fluid	
Rancher and 250EX models	DOT 4
Recon models..	DOT 3 or DOT 4
Miscellaneous	
Wheel bearings	Medium weight, lithium-based multi-purpose grease (NLGI no. 3)
Swingarm pivot bearings	Medium weight, lithium-based multi-purpose grease (NLGI no. 3)
Cables and lever pivots	Chain and cable lubricant or 10W30 motor oil
Brake pedal/shift lever/throttle lever pivots	Chain and cable lubricant or 10W30 motor oil

1 Honda TRX250/350
Routine maintenance schedule

Routine maintenance intervals

Note: *The pre-ride inspection outlined in the owner's manual covers checks and maintenance that should be carried out on a daily basis. It's condensed and included here to remind you of its importance. Always perform the pre-ride inspection at every maintenance interval (in addition to the procedures listed). The intervals listed below are the shortest intervals recommended by the manufacturer for each particular operation during the model years covered in this manual. Your owner's manual may have different intervals for your model.*

Daily or before riding

Check the engine oil level
Check the fuel level and inspect for leaks
Check the operation of both brakes - check the front brake fluid level and look for leakage; check the rear brake pedal and lever for correct freeplay
Check the tires for damage, the presence of foreign objects and correct air pressure
Check the throttle for smooth operation and correct freeplay
Make sure the steering operates smoothly
Check for proper operation of the headlight, taillight, indicator lights, speedometer and horn
Make sure the engine kill switch works properly
Check the driveaxle boots (4wd models) for damage or deterioration
Check the air cleaner drain tube and clean it if necessary
Make sure any cargo is properly loaded and securely fastened
Check all fasteners, including wheel nuts and axle nuts, for tightness
Check the underbody for mud or debris that could start a fire or interfere with vehicle operation

Every 600 miles/100 operating hours

Perform all of the daily checks plus:
Check front brake fluid level
Inspect the brakes
Check and adjust the valve clearances
Clean the air filter element (1)

Clean the air cleaner housing drain tube (2)
Check/adjust the idle speed
Change the engine oil and oil filter
Check the tightness of all fasteners
Inspect the suspension
Clean and gap the spark plug
Check/adjust the reverse selector cable freeplay
Check the skid plates for looseness or damage
Adjust the clutch
Check the exhaust system for leaks and check fastener tightness; clean the spark arrester
Inspect the wheels and tires

Every 1200 miles/200 operating hours

Clean the centrifugal oil filter (Recon and 250EX models)
Check/adjust the throttle lever freeplay
Check choke operation
Check the cleanliness of the fuel system and the condition of the fuel line
Clean the fuel tap strainer screen
Check the brake shoes for wear (1, 2)
Check differential oil level
Inspect the steering system and steering shaft bearing

Every two years

Change the brake fluid
Change the differential oil

1 *More often in dusty, sandy or snowy conditions.*
2 *More often in wet or muddy conditions.*

2   Introduction to tune-up and routine maintenance

Refer to illustration 2.1

This Chapter covers in detail the checks and procedures necessary for the tune-up and routine maintenance of your vehicle. Section 1 includes the routine maintenance schedule, which is designed to keep the machine in proper running condition and prevent possible problems. The remaining Sections contain detailed procedures for carrying out the items listed on the maintenance schedule, as well as additional maintenance information designed to increase reliability. Maintenance information is also printed on decals, which are mounted in various locations on the vehicle **(see illustration)**. Where information on the decals differs from that presented in this Chapter, use the decal information.

Since routine maintenance plays such an important role in the safe and efficient operation of your vehicle, it is presented here as a comprehensive check list. For the rider who does all of the maintenance, these lists outline the procedures and checks that should be done on a routine basis.

Deciding where to start or plug into the routine maintenance schedule depends on several factors. If you have a vehicle whose warranty has recently expired, and if it has been maintained according to the warranty standards, you may want to pick up routine maintenance as it coincides with the next mileage or calendar interval. If you have owned the machine for some time but have never performed any maintenance on it, then you may want to start at the nearest interval and include some additional procedures to ensure that nothing important is overlooked. If you have just had a major engine overhaul, then you may want to start the maintenance routine from the beginning. If you have a used machine and have no knowledge of its history or maintenance record, you may desire to combine all the checks into one large service initially and then settle into the maintenance schedule prescribed.

The Sections which actually outline the inspection and maintenance procedures are written as step-by-step comprehensive guides to the actual performance of the work. They explain in detail each of the routine inspections and maintenance procedures on the check list. References to additional information in applicable Chapters are also included and should not be overlooked.

Before beginning any actual maintenance or repair, the machine should be cleaned thoroughly, especially around the oil filter housing, spark plug, cylinder head cover, side covers, carburetor, etc. Cleaning will help ensure that dirt does not contaminate the engine and will allow you to detect wear and damage that could otherwise easily go unnoticed.

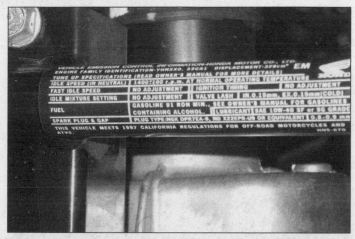

2.1 Decals on the vehicle include maintenance and safety information

3   Fluid levels - check

Engine oil

Refer to illustration 3.3a and 3.3b

1   Park the vehicle in a level position, then start the engine and allow it to reach normal operating temperature. **Caution:** *Do not run the engine in an enclosed space such as a garage or shop.*

2   Stop the engine and allow the machine to sit undisturbed in a level position for about five minutes.

3   With the engine off, unscrew the dipstick from the left side of the crankcase **(see illustrations)**. Pull it out, wipe it off with a clean rag, and reinsert it (let the dipstick rest on the threads; don't screw it back in). Pull the dipstick out and check the oil level on the dipstick scale. The oil level should be between the Maximum and Minimum level marks on the scale.

4   If the level is below the Minimum mark, add oil through the dipstick hole. Add enough oil of the recommended grade and type to bring the level up to the Maximum mark. Do not overfill.

Brake fluid

Refer to illustration 3.7

5   In order to ensure proper operation of the hydraulic front brakes,

3.3a The engine oil level must be between the upper and lower marks on the dipstick (here's the 350 dipstick) . . .

3.3b . . . and here's the dipstick on 250 models

3.7 The brake fluid level must be above the Lower mark on the reservoir; remove the cover screws (arrows) to add fluid

3.15 Unscrew the filler plug (upper arrow) to check front differential oil level; unscrew the drain bolt (hidden, lower arrow) to change the oil

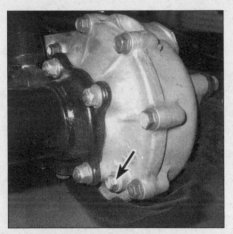

3.16a Unscrew the rear differential check bolt (arrow); if oil doesn't run out . . .

3.16b . . . unscrew the filler cap (arrow) and add oil until it runs out the check bolt; this is a Rancher . . .

3.16c . . . and this is a Recon/TRX250EX

the fluid level in the master cylinder reservoir must be properly maintained.

6   With the vehicle parked in a level position, turn the handlebars until the top of the front brake master cylinder is as level as possible.

7   The fluid level is visible through the sight glass in the master cylinder reservoir. Make sure that the fluid level is above the Lower mark on the reservoir **(see illustration)**.

8   If the level is low, the fluid must be replenished. Before removing the master cylinder cap, place rags beneath the reservoir (to protect the paint from brake fluid spills) and remove all dust and dirt from the area around the cap.

9   Remove the cover screws, then lift off the cover, rubber diaphragm and float (if equipped). **Note:** *Don't operate the brake lever with the cover removed.*

10   Add new, clean brake fluid of the recommended type until the level is even with the cast line inside the master cylinder reservoir. Don't mix different brands of brake fluid in the reservoir, as they may not be compatible. Also, don't mix different specifications (DOT 3 with DOT 4).

11   Reinstall the float (if equipped), rubber diaphragm and cover. Tighten the cover screws securely.

12   Wipe any spilled fluid off the reservoir body.

13   If the brake fluid level was low in either circuit, inspect the front or rear brake system for leaks.

Differential oil

Refer to illustrations 3.15, 3.16a, 3.16b and 3.16c

14   Park the vehicle on a level surface.

15   If you're working on a front differential, remove the differential filler cap **(see illustration)**. Feel the oil level inside the differential; it should be up to the bottom of the filler hole threads. If not, add the recommended oil until it does.

16   If you're working on a rear differential, remove the oil level check bolt **(see illustration)**. Oil should flow from the hole. If it doesn't, remove the oil filler cap **(see illustrations)**. Slowly pour oil into the filler cap hole until it flows from the check bolt hole, then install the check bolt and filler cap.

17   Reinstall the filler cap and tighten securely.

4   Battery - check

Warning: *Be extremely careful when handling or working around the battery. The electrolyte is very caustic and an explosive gas (hydrogen) is given off when the battery is charging.*

Refer to illustrations 4.1a and 4.1b

1   Remove the seat (see Chapter 7). If you're working on a Rancher, remove the battery cover **(see illustration)**. If you're working on a

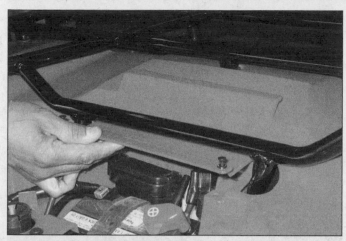

4.1a If you're working on a Rancher, remove the battery cover

4.1b On Recon models, unhook the rubber retaining band

Recon, unhook the retaining band **(see illustration)**. If you're working on a TRX250EX, remove the retainer bolts.

2 Remove the screws securing the battery cables to the battery terminals (remove the negative cable first, positive cable last). Lift the battery out of the vehicle.

3 The battery is a sealed type which requires no maintenance. **Note:** *Do not attempt to remove the battery caps to check the electrolyte level or battery specific gravity. Removal will damage the caps, resulting in electrolyte leakage and battery damage. All that should be done is to check that its terminals are clean and tight and that the casing is not damaged or leaking. See Chapter 8 for further details.*

4 If the vehicle will be stored for an extended time, fully charge the battery, then disconnect the negative cable before storage.

5 Install the battery. Be sure to refer to safety precautions regarding battery installation in Chapter 8.

5 Brake system - general check

1 A routine general check of the brakes will ensure that any problems are discovered and remedied before the rider's safety is jeopardized.

2 Check the brake levers and pedal for loose connections, excessive play, bends, and other damage. Replace any damaged parts with new ones (see Chapter 6).

3 Make sure all brake fasteners are tight. Check the brakes for wear

as described below and make sure the fluid level in the reservoir is correct (see Section 3). Look for leaks at the hose connections and check for cracks in the hoses. If the front brake lever is spongy, bleed the brakes as described in Chapter 6.

4 Make sure the brake light operates when the front brake lever is depressed. The front brake light switch is not adjustable. If it fails to operate properly, replace it with a new one (see Chapter 8).

5 Operate the rear brake lever and pedal. If operation is rough or sticky, refer to Section 12 and lubricate the cables.

Front brakes

Rancher and Recon models

Refer to illustration 5.6

6 Remove the adjusting hole plug from the brake drum **(see illustration)**.

7 Look through the hole to inspect the thickness of the lining material on the brake shoes (use a flashlight if necessary). If it's worn to near the limit listed in this Chapter's Specifications, refer to Chapter 6 and replace the brake shoes.

250EX models

Refer to illustration 5.8

8 Locate the wear indicator reference plate above the caliper **(see illustration)**. If the end of the indicator casting aligns with the plate, replace the brake pads (see Chapter 6).

5.6 On drum brake models, pull the rubber inspection plug out of the brake drum to check shoe wear

5.8 On TRX250EX models, replace the brake pads when the wear indicator reaches the limit line

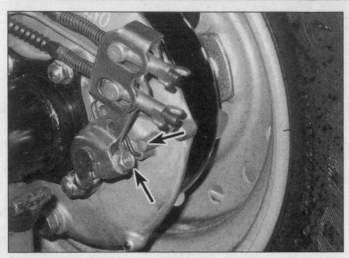

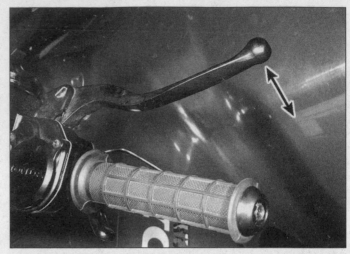

5.9 If the pointer (lower arrow) aligns with the mark (upper arrow) when the rear brake is applied, it's time to replace the rear brake shoes

6.1 Measure the freeplay of the front brake lever at the lever tip

Rear brakes

Refer to illustration 5.9

9   With the rear brake lever and pedal freeplay properly adjusted (see Section 6), check the wear indicator on the rear brake panel **(see illustration)**. If the pointer lines up with the indicator when the lever is pulled or the pedal is pressed, refer to Chapter 6 and replace the brake shoes.

---

6   Brake lever and pedal freeplay - check and adjustment

Front brake lever

Refer to illustrations 6.1, 6.4a and 6.4b

1   Squeeze the front brake lever and note how far the lever travels **(see illustration)**.

Rancher and Recon models

2   If lever travel exceeds the limit listed in this Chapter's Specifications, adjust the front brakes as described below.
3   Securely block the rear wheels so the vehicle can roll. Jack up the front end and support it securely on jackstands.
4   There's an adjuster wheel at each wheel cylinder, located at the

sides of the brake panel **(see illustrations)**. It's accessible through the adjusting hole plug **(see illustration 5.6)**.
5   To adjust the brakes, insert a screwdriver through the adjusting hole and turn the adjuster wheel until the tire can't be turned by hand, then back it off three notches. Spin the tire by hand to make sure the brake lining isn't dragging on the drum; if it is, back off the adjuster just enough so the dragging stops. Then align the hole with the second adjuster wheel and repeat the adjustment.
6   Push the adjusting hole cap securely into its hole with a screw-driver.
7   Repeat the adjustment on the other front wheel, then remove the jackstands and lower the vehicle.

250EX models

8   If lever travel exceeds the value listed in this Chapter's Specifications, bleed the brakes as described in Chapter 6. If that doesn't help, the caliper and master cylinder must be overhauled or replaced with new ones.

Rear brakes

Refer to illustrations 6.9, 6.10 and 6.11

9   Check the rear brake lever play at the left handlebar in the same way as for front brake lever play **(see illustration)**. If it exceeds the limit listed in this Chapter's Specifications, adjust it as described below.
10   If freeplay isn't within specifications, turn the lower wingnut at the

6.4a Insert a screwdriver through the access hole to turn the adjuster wheels . . .

6.4b . . . there's one on each wheel cylinder (brake drum removed for clarity)

6.9 Check the rear brake lever freeplay at the handlebar

6.10 The upper wingnut adjusts the lever; the lower wingnut adjusts the pedal

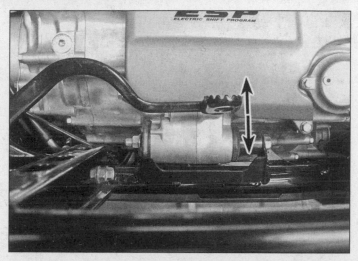

6.11 Measure freeplay at the pedal

brake panel lever (see illustration). Note: *Push the lever forward so its bushing clears the cutout in the wingnut, then turn the nut. Make sure the cutout seats on the bushing after adjustment.*

11 Check the play of the rear brake pedal (see illustration). If it exceeds the limit listed in this Chapter's Specifications, adjust it with the upper wingnut at the brake panel (see illustration 6.10).

7 Tires/wheels - general check

Refer to illustration 7.4

1 Routine tire and wheel checks should be made with the realization that your safety depends to a great extent on their condition.

2 Check the tires carefully for cuts, tears, embedded nails or other sharp objects and excessive wear. Operation of the vehicle with excessively worn tires is extremely hazardous, as traction and handling are directly affected. Measure the tread depth at the center of the tire and replace worn tires with new ones when the tread depth is less than that listed in this Chapter's Specifications.

3 Repair or replace punctured tires as soon as damage is noted. Do not try to patch a torn tire, as wheel balance and tire reliability may be impaired.

4 Check the tire pressures when the tires are cold and keep them properly inflated (see illustration). Proper air pressure will increase tire life and provide maximum stability and ride comfort. Keep in mind that low tire pressures may cause the tire to slip on the rim or come off, while high tire pressures will cause abnormal tread wear and unsafe handling.

5 The steel wheels used on this machine are virtually maintenance free, but they should be kept clean and checked periodically for cracks, bending and rust. Never attempt to repair damaged wheels; they must be replaced with new ones.

6 Check the valve stem locknuts to make sure they're tight. Also, make sure the valve stem cap is in place and tight. If it is missing, install a new one made of metal or hard plastic.

7.4 Check tire pressure with a gauge that will read accurately at the low pressures used in ATV tires

8 Reverse lock system - check and adjustment

Refer to illustrations 8.2, 8.3 and 8.4

All except 2006 and later TRX250EX

1 Follow the reverse selector cable from its lever on the left handlebar to its lever on the right side of the engine. Check for kinks, bends, loose retainers or other problems and correct them as necessary.

2 Measure the gap between the reverse lock lever and its cable bracket at the handlebar (see illustration). If it's not within the range listed in this Chapter's Specifications, adjust it.

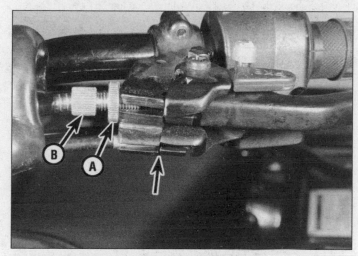

8.2 On all except 2006 and later 250EX models, this lever controls reverse selection and its freeplay is measured at the gap (arrow) - on 2006 and later 250EX models, the lever controls the clutch and its freeplay is measured at the lever tip

A Lockwheel   B Adjuster

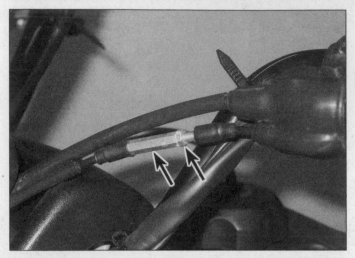

8.3 Loosen the locknut (right arrow) and turn the adjuster (left arrow) to adjust freeplay

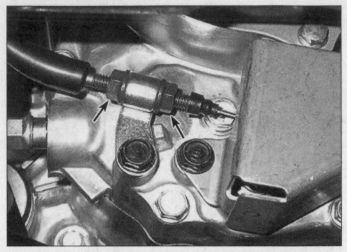

8.4 On 250 models (except the 2006 and later TRX250EX), loosen the locknut and turn the adjusting nut (arrows)

3   To adjust the cable on Rancher models, loosen the locknut at the handlebar adjuster **(see illustration)**. Turn the adjusting nut to achieve the correct play at the handlebar, then tighten the locknut securely.

4   To adjust the cable on Recon and 2005 and earlier 250EX models, loosen the locknut at the cable adjuster **(see illustration)**. Turn the adjusting nut to achieve the correct play at the handlebar, then tighten the locknut securely.

2008 and later TRX250EX

5   The left handlebar lever on these models controls the clutch rather than the reverse selector as on earlier models, so reverse selection is done by a knob on the frame.

6   If you're working on a 2006 or 2007 model, operate the knob and note its freeplay. If it's not within the range listed in this Chapter's Specifications, follow the cable down to the engine. Loosen the locknut on the cable and turn the adjusting nut to change freeplay at the knob. If freeplay still can't be adjusted, the cable may be stretched or the clutch itself may be worn.

7   If you're working on a 2008 or later model, turn the knob to the counterclockwise position and try to shift into reverse with the engine running. This should not be possible. With the knob turned fully clockwise, it should be possible to shift into reverse. If the reverse selector doesn't perform as described, check the cable for damage or incorrect routing.

9   Clutch - check and freeplay adjustment

Refer to illustrations 9.2a and 9.2b

All except 2006 and later Recon

1   The automatic clutch mechanism on these models disengages the change clutch automatically when the shift lever is operated, so there is no clutch lever (as is normal on motorcycles). If shifting gears becomes difficult, the clutch may be in need of adjustment.

2   Loosen the locknut on the front of the engine **(see illustration)**. Carefully turn the adjusting screw counterclockwise until you feel resistance, then turn it back in 1/4 turn. Hold the screw in this position and tighten the locknut to the torque listed in this Chapter's Specifications.

2008 and later TRX250EX

3   These vehicles are equipped with a cable-type clutch, operated by a lever on the left handlebar. Operate the lever and note how much freeplay there is (how much the lever moves before resistance suddenly increases). Compare this to the value listed in this Chapter's Specifications and adjust the clutch if freeplay is not within the specified range.

4   To adjust, pull back the rubber cover at the handlebar lever for access to the adjuster **(see illustration 8.2)**. Loosen the lockwheel (the larger nut on the cable) and turn the adjuster to make minor adjust-

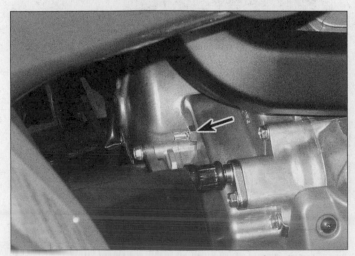

9.2a Loosen the locknut and turn the screw as described in the text, then tighten the locknut - here's the 350 adjuster . . .

9.2b . . . and here's the adjuster on 250 models

10.1 Check the driveaxle boots (arrows) for damage or deterioration

11.2 Measure throttle freeplay at the lever

11.3 Loosen the lockwheel (right arrow) and turn the adjuster wheel (left arrow) to make fine adjustments in throttle freeplay

11.5 Here are the carburetor throttle cable locknut (A), adjuster (B) and choke cable (C) (350 shown; 250 similar)

ments. Once the freeplay is set correctly, tighten the lockwheel and slip the cover back into position.
5   If the cable can't be adjusted at the handlebar, follow it down to the engine. Loosen the locknut on the cable and turn the adjusting nut to change freeplay at the lever. If freeplay still can't be adjusted, the cable may be stretched or the clutch itself may be worn.

10   Driveaxle boots (4wd models) - check

Refer to illustration 10.1
1   Check the driveaxle boots for cracks or damage, such as tears and cuts **(see illustration)**.
2   If any problems are found, refer to Chapter 5 and replace the boots.

11   Throttle operation/grip freeplay - check and adjustment

Throttle check
Refer to illustration 11.2
1   Make sure the throttle lever moves easily from fully closed to fully open with the front wheel turned at various angles. The grip should

return automatically from fully open to fully closed when released. If the throttle sticks, check the throttle cable for cracks or kinks in the housings. Also, make sure the inner cable is clean and well-lubricated.
2   Check for a small amount of freeplay at the lever and compare the freeplay to the value listed in this Chapter's Specifications **(see illustration)**.

Throttle adjustment
Refer to illustration 11.3
3   Freeplay adjustments can be made at the throttle lever end of the accelerator cable. Pull back the rubber boot and loosen the lockwheel on the cable **(see illustration)**. Turn the adjuster until the desired freeplay is obtained, then retighten the lockwheel.

Rancher models
Refer to illustration 11.5
4   If the freeplay can't be adjusted at the grip end, adjust the cable at the carburetor end. To do this, first remove the seat (see Chapter 7).
5   Loosen the locknut on the throttle cable **(see illustration)**. Turn the adjusting nut to set freeplay, then tighten the locknut securely.

12.2 On 250 models, make sure the choke lever (arrow) moves freely

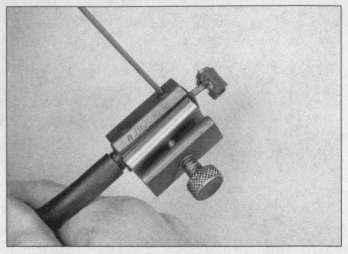

13.3 Lubricating a cable with a pressure lube adapter (make sure the tool seats around the inner cable)

12   Choke - operation check

Refer to illustration 12.2

1   Operate the choke knob on the left handlebar while you feel for smooth operation. If the lever doesn't move smoothly, refer to Section 13 and lubricate the choke cable.
2   Follow the cable from the handlebar to the choke lever or starting enrichment valve on the engine **(see illustration 11.5 or the accompanying illustration)**. Check for kinks, bends, loose retainers or other problems and correct them as necessary.

13   Lubrication - general

Refer to illustration 13.3

1   Since the controls, cables and various other components of a vehicle are exposed to the elements, they should be lubricated periodically to ensure safe and trouble-free operation.
2   The throttle and brake levers, brake pedal, kickstarter pivot should be lubricated frequently. In order for the lubricant to be applied where it will do the most good, the component should be disassembled. However, if chain and cable lubricant is being used, it can be applied to the pivot joint gaps and will usually work its way into the areas where friction occurs. If motor oil or light grease is being used, apply it sparingly as it may attract dirt (which could cause the controls to bind or wear at an accelerated rate). **Note:** *One of the best lubricants for the control lever pivots is a dry-film lubricant (available from many sources by different names).*
3   The throttle, choke, brake and reverse cables should be removed and treated with a commercially available cable lubricant which is specially formulated for use on vehicle control cables. Small adapters for pressure lubricating the cables with spray can lubricants are available and ensure that the cable is lubricated along its entire length **(see illustration)**. When attaching the cable to the lever, be sure to lubricate the barrel-shaped fitting at the end with multi-purpose grease.
4   To lubricate the cables, disconnect them at the lower end, then lubricate the cable with a pressure lube adapter **(see illustration 13.3)**. See Chapter 2 (reverse selector cable), Chapter 3 (throttle and choke cables) or Chapter 6 (brake cables).
5   Refer to Chapter 5 for the following lubrication procedures:

a) *Swingarm bearing and dust seals*
b) *Front driveaxle splines (4wd models)*
c) *Rear driveshaft pinion joint*
d) *Rear axle shaft splines*

6   Refer to Chapter 6 for the following lubrication procedures:

a) *Brake pedal pivot and seals*

b) *Front brake drum waterproof seals*
c) *Rear brake drum cover seal*

14   Engine oil/filter - change

1   Consistent routine oil and filter changes are the single most important maintenance procedure you can perform on a vehicle. The oil not only lubricates the internal parts of the engine, transmission and clutch, but it also acts as a coolant, a cleaner, a sealant, and a protectant. Because of these demands, the oil takes a terrific amount of abuse and should be replaced often with new oil of the recommended grade and type. Saving a little money on the difference in cost between a good oil and a cheap oil won't pay off if the engine is damaged. Honda recommends against using the following:

a) *Oils with graphite or molybdenum additives*
b) *Non-detergent oils*
c) *Castor or vegetable based oils*
d) *Oil additives.*

2   Before changing the oil and filter, warm up the engine so the oil will drain easily. Be careful when draining the oil, as the exhaust pipe, the engine and the oil itself can cause severe burns.
3   Park the vehicle over a clean drain pan.
4   Remove the dipstick/oil filler cap to vent the crankcase and act as a reminder that there is no oil in the engine.

Rancher models

Refer to illustrations 14.5, 14.6a, 14.6b and 14.6c

5   Remove the drain plug from the engine **(see illustration)** and allow the oil to drain into the pan. Do not lose the sealing washer on the drain plug.
6   As the oil is draining, remove the oil filter cover bolts, then remove the cover, filter element and spring **(see illustrations)**. If additional maintenance is planned for this time period, check or service another component while the oil is allowed to drain completely.
7   Wipe any remaining oil out of the filter housing area of the crankcase.
8   Check the condition of the drain plug threads and the sealing washer.
9   Install the spring on its post, then install the filter element **(see illustrations 14.6c and 14.6b)**. **Caution:** *The OUTSIDE mark on the filter must face outward (away from the crankcase) or severe engine damage will occur.*
10   Install new O-rings on the crankcase and on the filter cover **(see illustrations 14.6c and 14.6b)**. Install the cover and tighten its bolts to the torque listed in this Chapter's Specifications.

14.5 The oil drain plug on 350 models is on the underside of the crankcase

14.6a Remove the filter cover bolts (arrows) and lift off the cover

Recon and 250EX models

Refer to illustration 14.11

11   Remove the drain plug from the engine **(see illustration)** and allow the oil to drain into the pan. Do not lose the sealing washer on the drain plug.

12   If you're just changing the oil, skip to Step 14.

13   **Note:** *This step is somewhat complicated and can be time consuming. Before starting, read through the procedures referred to in Chapter 2.* At alternate oil changes, clean the centrifugal oil filter. To do this, remove the front crankcase cover and unbolt the outer clutch cover (also called the centrifugal filter cover) from the centrifugal clutch (see Chapter 2). Wipe the cover and centrifugal clutch with a lint-free cloth. Inspect the outer cover gasket or O-ring and install a new one if necessary. Refer to Chapter 2 and install all parts removed.

All models

14   Slip a new sealing washer over the drain plug, then install and tighten the plug to the torque listed in this Chapter's Specifications. Avoid overtightening, as damage to the engine case will result.

15   Before refilling the engine, check the old oil carefully. If the oil was drained into a clean pan, small pieces of metal or other material can be easily detected. If the oil is very metallic colored, then the engine is experiencing wear from break-in (new engine) or from insufficient lubri-

14.6b Pull off the cover and note the O-ring locations (arrows); be sure the filter element's OUT SIDE mark faces outward on installation

14.6c Remove the spring from its post and wipe the remaining oil out of the filter housing

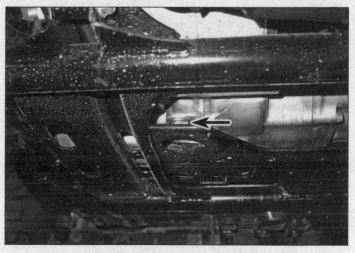

14.11 Here's the engine oil drain plug on 250 models

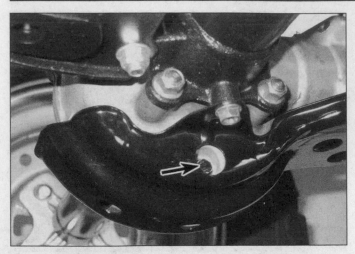

15.3 Here's the differential drain bolt on 250 models

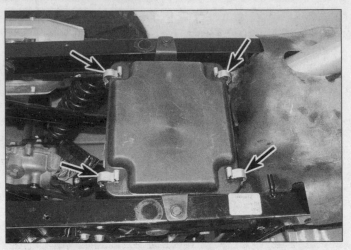

16.2 Pull back the clips (arrows) and lift off the cover

cation. If there are flakes or chips of metal in the oil, then something is drastically wrong internally and the engine will have to be disassembled for inspection and repair.

16   If there are pieces of fiber-like material in the oil, the change clutch is experiencing excessive wear and should be checked.

17   If the inspection of the oil turns up nothing unusual, refill the crankcase to the proper level with the recommended oil and install the dipstick/filler cap. Start the engine and let it run for two or three minutes. Shut it off, wait a few minutes, then check the oil level. If necessary, add more oil to bring the level up to the upper level mark on the dipstick. Check around the drain plug and filter cover or front crankcase cover for leaks.

18   The old oil drained from the engine cannot be reused in its present state and should be disposed of. Check with your local refuse disposal company, disposal facility or environmental agency to see whether they will accept the oil for recycling. Don't pour used oil into drains or onto the ground. After the oil has cooled, it can be drained into a suitable container (capped plastic jugs, topped bottles, milk cartons, etc.) for transport to one of these disposal sites.

15   Differential oil - change

Refer to illustration 15.3
1   If you're working on a 4wd model, place a protective cover under the front differential drain bolt so oil won't drip onto the frame rail. This

can be a piece of cardboard or aluminum foil.
2   Place a drain pan beneath the differential. **Note:** *It's a good idea to remove the differential skid plate (see Chapter 7), even though the drain bolt is accessible through a hole in the plate. If the plate is left on, oil will tend to spill on it, especially if there is debris trapped in the plate. This will then drip after the oil change is complete.*
3   Remove the oil filler plug, then the drain bolt and sealing washer **(see illustration 3.15 or the accompanying illustration).** Let the oil drain for several minutes, until it stops dripping.
4   Clean the drain bolt and sealing washer. If the sealing washer is in good condition, it can be reused; otherwise, replace it.
5   Install the drain bolt and tighten it to the torque listed in this Chapter's Specifications.
6   Add oil of the type and amount listed in this Chapter's Specifications, then check the oil level as described in Section 3. Install the filler plug and check bolt) and tighten it to the torque listed in this Chapter's Specifications.
7   Refer to Step 18 in Section 14 to dispose of the drained oil.

16   Air cleaner - filter element and drain tube cleaning

Element cleaning
Refer to illustrations 16.2, 16.3 and 16.4
1   Remove the seat (see Chapter 7).
2   Remove the clips that secure the filter cover and lift it off **(see illustration).**
3   Loosen the clamping band at the front of the element, remove the holder screw and lift the element out **(see illustration).**
4   Separate the foam element from the metal core **(see illustration).**
5   Clean the element and core in a high flash point solvent, squeeze the solvent out of the foam and let the core and element dry completely.
6   Soak the foam element in the amount and type of foam filter oil listed in this Chapter's Specifications, then squeeze it firmly to remove the excess oil.
7   Place the element on the core.
8   The remainder of installation is the reverse of the removal steps.

Drain tube cleaning
Refer to illustration 16.9
9   Check the drain tube for accumulated water and oil **(see illustration).** If oil or water has built up in the tube, squeeze its clamp, remove it from the air cleaner housing and clean it out. Install the drain tube on the housing and secure it with the clamp.

16.3 Loosen the clamping band and remove the holder screw (arrows), then separate the filter element from the housing and lift out the holder, element and core

16.4 Separate the foam element from the metal core

16.9 Open the clip and pull the drain tube (arrow) off its fitting

17 Fuel system - check and filter cleaning

Refer to illustrations 17.1, 17.5a, 17.5b, 17.9a and 17.9b

Warning: *Gasoline is extremely flammable, so take extra precautions when you work on any part of the fuel system. Don't smoke or allow open flames or bare light bulbs near the work area, and don't work in a garage where a gas-type appliance (such as a water heater or clothes dryer) is present. Since gasoline is carcinogenic, wear protective gloves and if you spill any fuel on your skin, rinse it off immediately with soap and water. When you perform any kind of work on the fuel system, wear safety glasses and have a fire extinguisher suitable for class B type fires (flammable liquids) on hand.*

1   Check the carburetor, fuel tank, the fuel tap and the line for leaks and evidence of damage **(see illustration)**.

2   If carburetor gaskets are leaking, the carburetor should be disassembled and rebuilt by referring to Chapter 3.

3   If the fuel tap is leaking, tightening the screws may help. If leakage persists, the tap should be disassembled and repaired or replaced with a new one.

4   If the fuel line is cracked or otherwise deteriorated, replace it with a new one.

5   Place the fuel tap lever in the Off position. Remove the fuel tap bolts and remove the strainer and O-ring **(see illustrations)**.

6   Clean the strainer. If it's torn or otherwise damaged, replace it.

7   Installation is the reverse of the removal steps, with the following additions:

17.1 Check the fuel line (arrow) for cracks, damage or deterioration and replace it if there are any problems

a) *Install a new O-ring and tighten the nut to the torque listed in this Chapter's Specifications.*

b) *Tighten the fuel tap bolts to the torque listed in this Chapter's Specifications.*

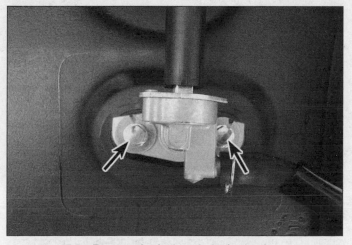

17.5a Remove the fuel tap bolts (arrows) . . .

17.5b . . . remove the fuel tap, then pull off the O-ring and plastic strainer

17.9a Here are the float chamber drain screw (left) and throttle stop screw (right) on 350 models

17.9b Here's the float chamber drain screw on 250 models

18.3 Unscrew the plug (arrow), then hold a rag against the muffler opening and rev the engine a few times to blow carbon out of the spark arrester

19.1 Twist the spark plug cap back and forth to free it, then pull it off the plug (350 shown)

8   After installation, run the engine and check for fuel leaks.

9   If the vehicle will be stored for a month or more, remove and drain the fuel tank. Also loosen the float chamber drain screw and drain the fuel from the carburetor **(see illustrations)**.

18   Exhaust system - inspection and spark arrester cleaning

Refer to illustration 18.3

1   Periodically check the exhaust system for leaks and loose fasteners. If tightening the holder nuts at the cylinder head fails to stop any leaks, replace the gasket with a new ones (a procedure which requires removal of the system).

2   The exhaust pipe flange nuts at the cylinder head are especially prone to loosening, which could cause damage to the head. Check them frequently and keep them tight.

3   **Warning:** *Make sure the exhaust system is cool before doing this procedure.* At the specified interval, remove the plate or plug from under the rear end of the muffler **(see illustration)**. Hold a rag firmly over the hole at the rear end of the muffler. Have an assistant start the engine and rev it a few times to blow out carbon, then shut the engine off.

4   After the exhaust system has cooled, install the plug. Tighten the plug securely.

19   Spark plug - replacement

Refer to illustrations 19.1, 19.5a and 19.5b

1   Twist the spark plug cap to break it free from the plug, then pull it off **(see illustration)**. If available, use compressed air to blow any accumulated debris from around the spark plug. Unscrew the plug with a spark plug socket.

2   Inspect the electrodes for wear. Both the center and side electrodes should have square edges and the side electrode should be of uniform thickness. Look for excessive deposits and evidence of a cracked or chipped insulator around the center electrode. Compare your spark plugs to the color spark plug reading chart. Check the threads, the washer and the ceramic insulator body for cracks and other damage.

3   If the electrodes are not excessively worn, and if the deposits can be easily removed with a wire brush, the plug can be regapped and reused (if no cracks or chips are visible in the insulator). If in doubt concerning the condition of the plug, replace it with a new one, as the expense is minimal.

19.5a Spark plug manufacturers recommend using a wire type gauge when checking the gap - if the wire doesn't slide between the electrodes with a slight drag, adjustment is required

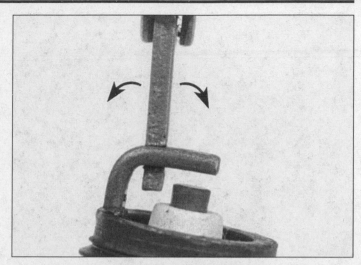

19.5b To change the gap, bend the side electrode only, as indicated by the arrows, and be very careful not to crack or chip the ceramic insulator surrounding the center electrode

4   Cleaning the spark plug by sandblasting is permitted, provided you clean the plug with a high flash-point solvent afterwards.

5   Before installing a new plug, make sure it is the correct type and heat range. Check the gap between the electrodes, as it is not preset. For best results, use a wire-type gauge rather than a flat gauge to check the gap **(see illustration)**. If the gap must be adjusted, bend the side electrode only and be very careful not to chip or crack the insulator nose **(see illustration)**. Make sure the washer is in place before installing the plug.

6   Since the cylinder head is made of aluminum, which is soft and easily damaged, thread the plug into the head by hand. Slip a short length of hose over the end of the plug to use as a tool to thread it into place. The hose will grip the plug well enough to turn it, but will start to slip if the plug begins to cross-thread in the hole - this will prevent damaged threads and the accompanying repair costs.

7   Once the plug is finger tight, the job can be finished with a socket. If a torque wrench is available, tighten the spark plug to the torque listed in this Chapter's Specifications. If you do not have a torque wrench, tighten the plug finger tight (until the washer bottoms on the cylinder head) then use a wrench to tighten it an additional 1/4 turn. Regardless of the method used, do not over-tighten it.

8   Reconnect the spark plug cap.

20 Valve clearances - check and adjustment

Refer to illustrations 20.5, 20.6, 20.9a and 20.9b

1   The engine must be completely cool for this maintenance procedure (below 35-degrees C/95-degrees F), so if possible let the machine sit overnight before beginning.

2   Refer to Section 4 and disconnect the cable from the negative terminal of the battery.

3   Refer to Chapter 3 and remove the fuel tank and heat shield.

4   Refer to Section 18 and remove the spark plug.

5   If you're working on a Rancher or a 1997 through 2001 Recon, remove the valve cover (see Chapter 2). If you're working on a 250EX or a 2002 or later Recon, unscrew the valve adjusting hole caps **(see illustration)**.

6   Unscrew the timing hole cap (just forward of the recoil starter pull rope) **(see illustration)**.

7   Position the piston at Top Dead Center (TDC) on the compression stroke. Do this by turning the crankshaft (pull on the recoil starter rope) until the mark on the rotor is aligned with the timing notch inside the hole **(see illustration 20.6)** and the piston is at the top of its travel. **Note:** *If the piston is at the top of its travel (look into the spark plug hole*

20.5 On all TRX250EX models and 2002 and later Recon models, unscrew the valve adjusting hole caps

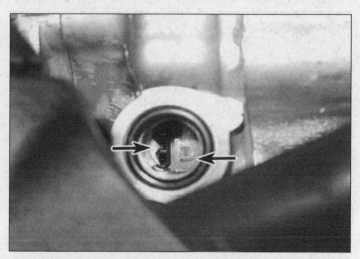

20.6 Align the pointer inside the crankcase cover (left arrow) with the TDC line on the alternator rotor inside the hole (right arrow) (350 shown)

20.9a Measure valve clearance with a feeler gauge; to adjust it, loosen the locknut and turn the adjusting screw with a screwdriver . . .

20.9b . . . on TRX250EX and 2002 and later Recon models, the adjusters are accessible through the hole caps

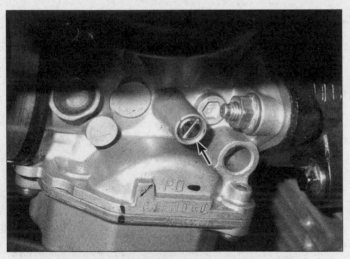

21.3 Here's the throttle stop screw on 250 models (arrow)

with a flashlight) and the mark is not visible in the timing hole, the piston is on the exhaust stroke rather than the compression stroke. Rotate the crankshaft one full turn so the timing mark and notch align.

8   With the engine in this position, both of the valves can be checked.

9   To check, insert a feeler gauge of the thickness listed in this Chapter's Specifications between the valve stem and rocker arm **(see illustrations)**. Pull the feeler gauge out slowly - you should feel a slight drag. If there's no drag, the clearance is too loose. If there's a heavy drag, the clearance is too tight.

10   If the clearance is incorrect, loosen the adjuster locknut with a box wrench. Turn the adjusting screw with a screwdriver (if it has a slot) or needle-nosed pliers (if it's square) until the correct clearance is achieved, then tighten the locknut. Note: On some models, it's necessary to insert the screwdriver through an access hole in the frame to adjust the intake valve (on the rear side of the engine).

11   After adjusting, recheck the clearance with the feeler gauge to make sure it wasn't changed when the locknut was tightened.

12   Now measure the other valve, following the same procedure you used for the first valve. Make sure to use a feeler gauge of the specified thickness.

13   With both of the clearances within the Specifications, install the valve adjusting hole covers and timing hole caps. Install the crankshaft hole cap (2wd) or reduction shaft hole cap (4wd). Use new O-rings on

the covers and caps if the old ones are hardened, deteriorated or damaged.

14   Install the fuel tank and reconnect the cable to the negative terminal of the battery.

21 Idle speed - check and adjustment

1   Before adjusting the idle speed, make sure the valve clearances and spark plug gap is correct. Also, turn the handlebars back-and-forth and see if the idle speed changes as this is done. If it does, the throttle cable may not be adjusted correctly, or it may be worn out. Be sure to correct this problem before proceeding.

2   The engine should be at normal operating temperature, which is usually reached after 10 to 15 minutes of stop and go riding. Make sure the transmission is in Neutral.

3   Turn the throttle stop screw **(see illustration 16.9 or the accompanying illustration)** until the idle speed listed in this Chapter's Specifications is obtained.

4   Snap the throttle open and shut a few times, then recheck the idle speed. If necessary, repeat the adjustment procedure.

5   If a smooth, steady idle can't be achieved, the fuel/air mixture may be incorrect. Refer to Chapter 3 for additional carburetor information.

22 Fasteners - check

1   Since vibration of the machine tends to loosen fasteners, all nuts, bolts, screws, etc. should be periodically checked for proper tightness. Also make sure all cotter pins or other safety fasteners are correctly installed.

2   Pay particular attention to the following:
   Spark plug
   Engine oil drain plug
   Oil filter cover bolts
   Gearshift lever
   Brake pedal
   Footpegs
   Engine mount bolts
   Shock absorber mount bolts
   Front axle nuts
   Rear axle nuts
   Skid plate bolts

3   If a torque wrench is available, use it along with the torque specifications at the beginning of this, or other, Chapters.

23 Suspension - check

1   The suspension components must be maintained in top operating condition to ensure rider safety. Loose, worn or damaged suspension parts decrease the vehicle's stability and control.

2   Lock the front brake and push on the handlebars to compress the front shock absorbers several times. See if they move up-and-down smoothly without binding. If binding is felt, the shocks should be disassembled and inspected as described in Chapter 5.

3   Check the tightness of all front suspension nuts and bolts to be sure none have worked loose.

4   Inspect the rear shock absorbers for fluid leakage and tightness of the mounting nuts and bolts. If leakage is found, the shock should be replaced.

5   Support the vehicle securely upright with its rear wheel off the ground. Grab the swingarm on each side, just ahead of the axle. Rock the swingarm from side to side - there should be no discernible movement at the rear. If there's a little movement or a slight clicking can be heard, make sure the swingarm pivot shafts are tight. If the pivot shafts are tight but movement is still noticeable, the swingarm will have to be removed and the bearings replaced as described in Chapter 5.

6   Inspect the tightness of the rear suspension nuts and bolts.

24 Steering system - inspection and toe-in adjustment

Inspection

1   This vehicle is equipped with a ball bearing at the lower end of the steering shaft and a hard rubber bushing at the upper end, which can become dented, rough or loose during normal use of the machine. In extreme cases, worn or loose parts can cause steering wobble that is potentially dangerous.

2   To check the bearings, block the rear wheels so the vehicle can't roll, jack up the front end and support it securely on jackstands.

3   Point the wheel straight ahead and slowly move the handlebars from side-to-side. Dents or roughness in the bearing or bushing will be felt and the bars will not move smoothly. **Note:** *Make sure any hesitation in movement is not being caused by the cables and wiring harnesses that run to the handlebars.*

4   If the handlebars don't move smoothly, or if they move horizontally, refer to Chapter 5 to remove and inspect the steering shaft bushing and bearing.

24.10 Hold and turn the tie rod by placing an open end wrench on the flats as shown; loosen the locknuts (arrows) so the tie rod can be turned

Toe-in adjustment

Refer to illustration and 24.10

5   Roll the vehicle forward onto a level surface and stop it with the front wheels pointing straight ahead.

6   Make a mark at the front and center of each tire, even with the centerline of the front hub.

7   Measure the distance between the marks with a toe-in gauge or steel tape measure.

8   Have an assistant push the vehicle backward while you watch the marks on the tires. Stop pushing when the tires have rotated exactly one-half turn, so the marks are at the backs of the tires.

9   Again, measure the distance between the marks. Subtract the front measurement from the rear measurement to get toe-in.

10   If toe-in is not as specified in this Chapter's Specifications, hold each tie-rod with a wrench on the flats and loosen the locknuts **(see illustration)**. Turn the tie-rods an equal amount to change toe-in. When toe-in is set correctly, tighten the locknuts to the torque listed in this Chapter's Specifications.

Notes

Chapter 2
Engine, clutch and transmission

Contents

	Section		Section
Alternator cover .... See Chapter 8		Oil pump and cooler - removal, inspection and installation	19
Cam chain tensioner - removal and installation	20	Operations possible with the engine in the frame	3
Camshaft, chain and sprockets - removal,		Operations requiring engine removal	4
inspection and installation	21	Piston - removal, inspection and installation	13
Centrifugal clutch - removal, inspection and installation	16	Piston rings - installation	14
Change clutch and release mechanism - removal,		Primary drive gear (350 models) - removal, inspection	
inspection and installation	17	and installation	18
Crankcase - disassembly and reassembly	25	Rear crankcase cover (350 models) .... See Chapter 8	
Crankcase components - inspection and servicing	26	Recoil starter - removal, inspection and installation	23
Crankshaft and balancer - removal, inspection and installation	28	Recommended break-in procedure	30
Cylinder - removal, inspection and installation	12	Reverse lock mechanism - cable replacement, removal,	
Cylinder compression - check	2	inspection and installation	22
Cylinder head and valves - disassembly, inspection		Rocker arms, pushrods and cylinder head - removal,	
and reassembly	11	inspection and installation	9
Engine - removal and installation	6	Transmission shafts and shift drum - removal, inspection	
Engine disassembly and reassembly - general information	7	and installation	27
External shift mechanism - removal, inspection and installation ..	24	Valve cover (1997 through 2001 Recon and all 350 models) -	
General information	1	removal, inspection and installation	8
Initial start-up after overhaul	29	Valve lifters - removal, inspection and installation	15
Major engine repair - general note	5	Valves/valve seats/valve guides - servicing	10

Specifications

350 models

General

Bore	78.5 mm (3.09 inches)
Stroke	68 mm (2.68 inches)
Displacement	329.1 cc (24.1 cubic inches)

Rocker arms and lifters

Rocker arm inside diameter	
Standard	12.000 to 12.018 mm (0.4724 to 0.4731 inch)
Limit	12.05 mm (0.474 inch)
Rocker shaft outside diameter	
Standard	11.966 to 11.984 mm (0.4711 to 0.4718 inch)
Limit	11.92 mm (0.469 inch)
Shaft-to-arm clearance	
Standard	0.016 to 0.052 mm (0.0006 to 0.0020 inch)
Limit	0.08 mm (0.003 inch)
Valve lifter diameter (intake and exhaust)	
Standard	22.467 to 22.482 mm (0.8845 to 0.8851 inch)
Limit	22.46 mm (0.884 inch)
Valve lifter bore diameter (intake and exhaust)	
Standard	22.510 to 22.526 mm (0.8862 to 0.8868 inch)
Limit	22.54 mm (0.887 inch)

350 models (continued)

Camshaft
Lobe height limit (intake and exhaust).. 35.13 mm (1.383 inch)

Cylinder head, valves and valve springs
Cylinder head warpage limit.. 0.10 mm 0.004 inch)
Valve stem runout.. not specified
Valve stem diameter
   Intake
      Standard .... 5.475 to 5.490 mm (0.2156 to 0.2161 inch)
      Limit .... 5.45 mm (0.215 inch)
   Exhaust
      Standard .... 5.455 to 5.470 mm (0.2148 to 0.2154 inch)
      Limit .... 5.43 mm (0.214 inch)
Valve guide inside diameter (intake and exhaust)
   Standard .... 5.500 to 5.512 mm (0.2165 to 0.2170 inch)
   Limit .... 5.52 mm (0.217 inch)
Valve seat width (intake and exhaust)
   Standard .... 1.2 mm (0.05 inch)
   Limit .... 1.5 mm (0.06 inch)
Valve spring free length
   Inner spring
      Standard .... 36.95 mm (1.455 inch)
      Limit .... 36.94 mm (1.454 inch)
   Outer spring
      Standard .... 41.67 mm (1.641 inch)
      Limit .... 40.42 mm (1.591 inch)

Cylinder
Bore diameter
   Standard .... 78.500 to 78.510 mm (3.0905 to 3.0909 inch)
   Limit .... 78.60 mm (3.094 inches)
Taper and out-of-round limits .... 0.10 mm (0.004 inch)
Surface warpage limit... 0.10 mm (0.004 inch)

Pistons
Piston diameter
   Standard .... 78.465 to 78.485 mm (3.0892 to 3.0900 inch)
   Limit .... 78.43 mm (3.088 inches)
Piston-to-cylinder clearance
   Standard .... 0.15 to 0.45 mm (0.0006 to 0.0018 inch)
   Limit .... 0.10 mm (0.004 inch)
Oversize pistons and rings available... + 0.25 mm (0.010 inch), 0.5 mm (0.020 inch), 0.75 mm (0.030 inch), 1.00 mm (0.040 inch)
Piston pin bore
   In piston
      Standard .... 17.002 to 17.008 mm (0.6694 to 0.6696 inch)
      Limit .... 17.04 mm (0.671 inch)
   In connecting rod
      Standard .... 17.016 to 17.034 mm (0.6699 to 0.6706 inch)
      Limit .... 17.10 mm (0.673 inch)
Piston pin outer diameter
   Standard .... 16.994 to 17.000 mm (0.6691 to 0.6693 inch)
   Limit .... 16.96 mm (0.668 inch)
Piston pin-to-piston clearance
   Standard .... 0.002 to 0.014 mm (0.0001 to 0.0006 inch)
   Limit .... 0.02 mm (0.001 inch)
Ring side clearance
   Top
      Standard .... 0.030 to 0.060 mm (0.0012 to 0.0024 inch)
      Limit .... 0.09 mm (0.004 inch)
   Second
      Standard .... 0.015 to 0.045 mm (0.0006 to 0.0018 inch)
      Limit .... 0.09 mm (0.004 inch)
Ring end gap
   Top
      Standard .... 0.15 to 0.30 mm (0.006 to 0.012 inch)
      Limit .... 0.5 mm (0.02 inch)

Ring end gap
  Second
    Standard .... 0.30 to 0.45 mm (0.012 to 0.018 inch)
    Limit .... 0.60 mm (0.02 inch)
  Oil
    Standard .... 0.20 to 0.70 mm (0.008 to 0.028 inch)
    Limit .... 0.9 mm (0.04 inch)

Centrifugal clutch
Drum internal diameter
  Standard .... 126.0 to 126.2 mm (4.96 to 4.97 inches)
  Limit .... 126.4 mm (4.98 inches)
Weight lining thickness
  Standard .... 2.0 mm (0.08 inch)
  Limit .... 1.3 mm (0.05 inches)
Clutch spring height
  Standard .... 2.87 mm (0.113 inch)
  Limit .... 2.73 mm (0.107 inch)
Clutch weight spring free length
  Standard .... 25.8 mm (1.02 inch)
  Limit .... 26.9 mm (1.06 inch)

Change clutch
Spring free length
  Foot shift models
    Standard .... 28.0 mm (1.10 inch)
    Limit .... 27.0 mm (1.06 inch)
  Electric shift models
    Standard .... 31.3 mm (1.23 inch)
    Limit .... 30.2 mm (1.19 inch)
Friction plate thickness
  Standard .... 2.62 to 2.78 mm (0.103 to 0.109 inch)
  Limit .... 2.3 mm (0.09 inch)
Friction and metal plate warpage limit .... 0.20 mm (0.008 inch)
Clutch outer guide outside diameter
  Standard .... 27.959 to 27.980 mm (1.1007 to 1.1016 inch)
  Limit .... 27.92 mm (1.099 inch)
Clutch outer guide inside diameter
  Standard .... 22.000 to 22.021 mm (0.8661 to 0.8670 inch)
  Limit .... 22.05 mm (0.868 inch)
Diameter of outer guide friction surface on mainshaft
  Standard .... 21.967 to 21.980 mm (0.8648 to 0.8654 inch)
  Limit .... 21.93 mm (0.863 inch)
Primary drive gear
  Inside diameter
    Standard .... 27.000 to 27.021 mm (1.0630 to 1.0638 inch)
    Limit .... 27.05 mm (1.065 inch)
Diameter of friction surface on crankshaft
  Standard .... 26.959 to 26.980 mm (1.0614 to 1.0622 inch)
  Limit .... 26.93 mm (1.060 inch)

Oil pump
Outer rotor to body clearance
  Standard .... 0.15 to 0.22 mm (0.006 to 0.009 inch)
  Limit .... 0.25 mm (0.010 inch)
Inner to outer rotor clearance
  Standard .... 0.15 mm (0.006 inch) or less
  Limit .... 0.20 mm (0.008 inch)
Side clearance (rotors to straightedge)
  Standard .... 0.02 to 0.09 mm (0.001 to 0.004 inch)
  Limit .... 0.12 mm (0.006 inch)

Shift drum and forks
Fork inside diameter
  Standard .... 13.000 to 13.018 mm (0.5118 to 0.5125 inch)
  Limit .... 13.04 mm (0.513 inch)
Fork shaft outside diameter
  Standard .... 12.966 to 12.984 mm (0.5105 to 0.5112 inch)
  Limit .... 12.96 mm (0.510 inch)
Fork finger thickness
  Standard .... 4.93 to 5.00 mm (0.194 to 0.197 inch)
  Limit .... 4.50 mm (0.177 inch)

350 models (continued)

Transmission

Gear inside diameters

  Mainshaft fourth
    Standard .... 23.000 to 23.021 mm (0.9055 to 0.9063 inch)
    Limit .... 23.04 mm (0.907 inch)

  Mainshaft fifth
    Standard .... 18.000 to 18.021 mm (0.7087 to 0.7095 inch)
    Limit .... 18.05 mm (0.711 inch)

  Countershaft first, second, third, reverse
    Standard .... 25.000 to 25.021 mm (0.9843 to 0.9851 inch)
    Limit .... 25.05 mm (0.986 inch)

Reverse idler
    Standard .... 13.000 to 13.018 mm (0.5118 to 0.5125 inch)
    Limit .... 13.04 mm (0.513 inch)

Bushing diameters

  Mainshaft fourth outside diameter
    Standard .... 22.959 to 22.979 mm (0.9039 to 0.9047 inch)
    Limit .... 22.94 mm (0.903 inch)

  Mainshaft fourth inside diameter
    Standard .... 20.000 to 20.021 mm (0.7874 to 0.7882 inch)
    Limit .... 20.04 mm (0.789 inch)

  Mainshaft fifth outside diameter
    Standard .... 17.959 to 17.980 mm (0.7070 to 0.7079 inch)
    Limit .... 17.94 mm (0.706 mm)

  Mainshaft fifth inside diameter
    Standard .... 15.000 to 15.018 mm (0.5906 to 0.5913 inch)
    Limit .... 15.04 mm (0.592 inch)

  Countershaft first, second, third, reverse outside diameter
    Standard .... 24.959 to 24.980 mm (0.9826 to 0.9835 inch)
    Limit .... 24.93 mm (0.981 inch)

Gear-to-bushing clearances

  Mainshaft fourth
    Standard .... 0.021 to 0.062 mm (0.0008 to 0.0024 inch)
    Limit .... 0.10 mm (0.004 inch)

  Mainshaft fifth, countershaft first, second, third, reverse
    Standard .... 0.020 to 0.062 mm (0.0008 to 0.0024 inch)
    Limit .... 0.10 mm (0.004 inch)

Reverse idler gear-to-shaft clearance
    Standard .... 0.016 to 0.052 mm (0.0006 to 0.0020 inch)
    Limit .... 0.10 mm (0.004 inch)

Shaft diameters

  Mainshaft fourth gear surface
    Standard .... 19.959 to 19.980 mm (0.7858 to 0.7866 inch)
    Limit .... 19.93 mm (0.785 inch)

  Mainshaft fifth gear surface
    Standard .... 14.966 to 14.984 mm (0.5892 to 0.5899 inch)
    Limit .... 14.94 mm (0.588 inch)

  Reverse idler gear surface
    Standard .... 12.966 to 12.984 mm (0.5105 to 0.5112 inch)
    Limit .... 12.94 mm (0.509 inch)

Shaft-to-bushing clearances

  Mainshaft fourth, countershaft third
    Standard .... 0.020 to 0.062 mm (0.0008 to 0.0024 inch)
    Limit .... 0.10 mm (0.004 inch)

  Mainshaft fifth
    Standard .... 0.016 to 0.052 mm (0.0006 to 0.0020 inch)
    Limit .... 0.10 mm (0.004 inch)

Reverse idler shaft-to-gear clearance
    Standard .... 0.016 to 0.052 mm (0.0006 to 0.0020 inch)
    Limit .... 0.10 mm (0.004 inch)

Crankshaft

Connecting rod side clearance
    Standard .... 0.05 to 0.65 mm (0.002 to 0.026 inch)
    Limit .... 0.80 mm (0.031 inch)

Connecting rod big end radial clearance
    Standard .... 0.006 to 0.018 mm (0.0002 to 0.0007 inch)
    Limit .... 0.05 mm (0.002 inch)
Runout limit .... 0.05 mm (0.002 inch)

Torque specifications

Valve cover bolts	Not specified
Cylinder head cap nuts	39 Nm (29 ft-lbs) (1)
Rocker assembly holder bolt	30 Nm (22 ft-lbs)
Rocker shaft retaining bolt	7 Nm (61 inch-lbs)
Cam chain tensioner slider pivot bolt	12 Nm (108 inch-lbs)
Front crankcase cover bolts	Not specified
Centrifugal clutch locknut	118 Nm (87 ft-lbs) (2)
Change clutch spring bolts	Not specified
Change clutch locknut	108 Nm (80 ft-lbs) (2)
Oil pump	
To-engine bolts	Not specified
Cover screw	Not specified
Shift pedal pinch bolt	
2000 through 2003	Not specified
2004 and later	20 Nm (14 ft-lbs)
External shift linkage return spring pin	22 Nm (12 ft-lbs) (3)
Gearshift cam bolt	23 Nm (17 ft-lbs) (3)
Stopper arm bolt	12 Nm (108 inch-lbs)
Crankcase bolts	Not specified
Cylinder studs to crankcase	12 Nm (108 inch-lbs)
Engine upper hanger to frame nuts	54 Nm (40 ft-lbs)
Engine upper hanger to engine bolts	32 Nm (24 ft-lbs)
Lower engine hanger nuts (both sides)	54 Nm (40 ft-lbs)
Left engine hanger adjusting bolt	30 Nm (22 ft-lbs)

1. Use engine oil on the threads and the seating surfaces of the nuts.
2. Use a new locknut with engine oil on the threads and the seating surface of the nut. Stake the locknut.
3. Use non-permanent thread locking agent on the threads.

250 models

General

Bore	68.5 mm (2.70 inches)
Stroke	62.2 mm (2.45 inches)
Displacement	229.1 cc (14.0 cubic inches)

Rocker arms and lifters

Rocker arm inside diameter	
Standard	12.000 to 12.018 mm (0.4724 to 0.4731 inch)
Limit	12.05 mm (0.474 inch)
Rocker shaft outside diameter	
Standard	11.964 to 11.984 mm (0.4710 to 0.4718 inch)
Limit	11.92 mm (0.469 inch)
Shaft-to-arm clearance	
Standard	0.016 to 0.054 mm (0.0006 to 0.0021 inch)
Limit	0.08 mm (0.003 inch)
Valve lifter diameter (intake and exhaust)	
Standard	22.467 to 22.482 mm (0.8845 to 0.8851 inch)
Limit	22.46 mm (0.884 inch)
Valve lifter bore diameter (intake and exhaust)	
Standard	22.510 to 22.526 mm (0.8862 to 0.8868 inch)
Limit	22.54 mm (0.887 inch)

Camshaft

Lobe height limit	
Recon	
Intake	35.2 mm (1.39 inches)
Exhaust	35.0 mm (1.38 inches)
TRX250EX	
Intake	35.6 mm (1.40 inches)
Exhaust	35.1 mm (1.38 inches)

Cylinder head, valves and valve springs

Cylinder head warpage limit	0.10 mm (0.004 inch)
Valve stem runout	Not specified

250 models (continued)

Valve stem diameter
  Intake
    Standard .... 5.475 to 5.490 mm (0.2156 to 0.2161 inch)
    Limit .... 5.45 mm (0.215 inch)
  Exhaust
    Standard .... 5.455 to 5.470 mm (0.2148 to 0.2154 inch)
    Limit .... 5.43 mm (0.214 inch)
Valve guide inside diameter (intake and exhaust)
  Standard .... 5.500 to 5.512 mm (0.2165 to 0.2170 inch)
  Limit .... 5.52 mm (0.217 inch)
Valve seat width (intake and exhaust)
  Standard .... 1.2 mm (0.05 inch)
  Limit .... 1.5 mm (0.06 inch)
Valve spring free length
  2004 and earlier
    Inner spring
      Standard .... 36.95 mm (1.455 inches)
      Limit .... 35.7 mm (1.41 inches)
    Outer spring
      Standard .... 41.01 mm (1.615 inches)
      Limit .... 39.8 mm (1.57 inches)
  2005 and later
    Inner spring
      Standard .... 42.4 mm (1.67 inches)
      Limit .... 41.1 mm (1.62 inches)
    Outer spring
      Standard .... 44.2 mm (1.74 inches)
      Limit .... 42.9 mm (1.69 inches)

Cylinder

Bore diameter
  Standard .... 68.500 to 68.510 mm (2.6968 to 2.6972 inches)
  Limit .... 68.60 mm (2.70 inches)
Taper and out-of-round limits .... 0.10 mm (0.004 inch)
Surface warpage limit... 0.10 mm (0.004 inch)

Pistons

Piston diameter
  Standard .... 68.462 to 68.482 mm (2.6953 to 2.6961 inches)
  Limit .... 68.4 mm (2.69 inches)
Piston-to-cylinder clearance
  Standard .... 0.18 to 0.48 mm (0.0007 to 0.0019 inch)
  Limit .... 0.10 mm (0.004 inch)
Oversize pistons and rings available.................................. + 0.25 mm (0.010 inch), 0.5 mm (0.020 inch), 0.75 mm (0.030 inch),
                                      1.00 mm (0.040 inch)
Piston pin bore
  In piston
    Standard .... 15.002 to 15.008 mm (0.5906 to 0.5909 inch)
    Limit .... 15.04 mm (0.592 inch)
  In connecting rod
    Standard .... 15.010 to 15.028 mm (0.5909 to 0.59917 inch)
    Limit .... 15.06 mm (0.593 inch)
Piston pin outer diameter
  Standard .... 14.994 to 15.000 mm (0.5903 to 0.5906 inch)
  Limit .... 14.96 mm (0.589 inch)
Piston pin-to-piston clearance
  Standard .... 0.002 to 0.014 mm (0.0001 to 0.0006 inch)
  Limit .... 0.02 mm (0.001 inch)
Ring side clearance
  Top and second
    Standard .... 0.015 to 0.045 mm (0.0006 to 0.0018 inch)
    Limit .... 0.09 mm (0.004 inch)
Ring end gap
  Top
    Standard .... 0.20 to 0.35 mm (0.008 to 0.014 inch)
    Limit .... 0.5 mm (0.02 inch)

Second
    Standard... 0.40 to 0.55 mm (0.016 to 0.022 inch)
    Limit... 0.70 mm (0.03 inch)
Oil
    Standard... 0.20 to 0.70 mm (0.008 to 0.028 inch)
    Limit... Not specified

Centrifugal clutch

Drum internal diameter
    Standard... 116.0 to 116.2 mm (4.567 to 4.575 inches)
    Limit... 116.5 mm (4.59 inches)
Weight lining thickness
    Standard... 2.0 mm (0.08 inch)
    Limit... 1.2 mm (0.05 inch)
Clutch spring height
    Standard... 3.0 mm (0.12 inch)
    Limit... 2.85 mm (0.112 inch)
Clutch weight spring free length
    Standard... 30.75 mm (1.211 inches)
    Limit... 31.6 mm (1.24 inches)

Change clutch

Spring free length
    Recon
        All 2004 and earlier, 2005 and later foot shift
            Standard... 35.2 mm (1.39 inches)
            Limit... 34.5 mm (1.36 inches)
        2005 and later electric shift
            Standard... 37.0 mm (1.46 inches)
            Limit... 36.3 mm (1.43 inches)
    TRX250EX
        2005 and earlier
            Standard... 35.2 mm (1.39 inches)
            Limit... 34.5 mm (1.36 inches)
        2006 and later
            Standard... 38.9 mm (1.53 inches)
            Limit... 36.0 mm (1.417 inches)
Friction plate thickness
    Standard... 2.9 to 3.0 mm (0.11 to 0.12 inch)
    Limit .... 2.6 mm (0.10 inch)
Friction and metal plate warpage limit .... 0.20 mm (0.008 inch)
Clutch outer guide outside diameter
    Standard... 27.959 to 27.980 mm (1.1007 to 1.1016 inches)
    Limit .... 27.92 mm (1.099 inches)
Clutch outer guide inside diameter
    Standard... 28.000 to 28.021 mm (1.1024 to 1.1032 inches)
    Limit .... 28.05 mm (1.104 inch)
Oil pump
Outer rotor to body clearance
    Standard... 0.15 to 0.21 mm (0.006 to 0.008 inch)
    Limit .... 0.25 mm (0.010 inch)
Inner to outer rotor clearance
    Standard... 0.15 mm (0.006 inch) or less
    Limit .... 0.20 mm (0.008 inch)
Side clearance (rotors to straightedge)
    Standard... 0.05 to 0.13 mm (0.002 to 0.005 inch)
    Limit .... 0.15 mm (0.006 inch)

Shift drum and forks

Fork inside diameter
    Standard... 13.000 to 13.018 mm (0.5118 to 0.5125 inch)
    Limit .... 13.04 mm (0.513 inch)
Fork shaft outside diameter
    Standard... 12.966 to 12.984 mm (0.5105 to 0.5112 inch)
    Limit .... 12.96 mm (0.510 inch)
Fork finger thickness
    Standard... 4.93 to 5.00 mm (0.194 to 0.197 inch)
    Limit .... 4.50 mm (0.177 inch)

250 models (continued)

Transmission

Gear inside diameters
  Mainshaft fourth
    Standard .... 23.000 to 23.021 mm (0.9055 to 0.9063 inch)
    Limit .... 23.04 mm (0.907 inch)
  Mainshaft fifth
    Standard .... 18.000 to 18.021 mm (0.7087 to 0.7095 inch)
    Limit .... 18.04 mm (0.710 inch)
  Countershaft first, second, third, reverse
    Standard .... 25.000 to 25.021 mm (0.9843 to 0.9851 inch)
    Limit .... 28.05 mm (0.986 inch)
Reverse idler
  Standard .... 13.000 to 13.018 mm (0.5118 to 0.5125 inch)
  Limit .... 13.04 mm (0.513 inch)
Bushing diameters
  Mainshaft fourth outside diameter
    Standard .... 22.959 to 22.979 mm (0.9039 to 0.9047 inch)
    Limit .... 22.94 mm (0.903 inch)
  Mainshaft fourth inside diameter
    Standard .... 20.000 to 20.021 mm (0.7874 to 0.7882 inch)
    Limit .... 20.04 mm (0.789 inch)
  Mainshaft fifth outside diameter
    Standard .... 17.959 to 17.980 mm (0.7070 to 0.7079 inch)
    Limit .... 17.94 mm (0.706 mm)
  Mainshaft fifth inside diameter
    Standard .... 15.000 to 15.018 mm (0.5906 to 0.5913 inch)
    Limit .... 15.04 mm (0.592 inch)
  Countershaft first, second, third, reverse outside diameter
    Standard .... 24.959 to 24.980 mm (0.9826 to 0.9835 inch)
    Limit .... 24.94 mm (0.982 inch)
Gear-to-bushing clearances
  Mainshaft fourth
    Standard .... 0.021 to 0.062 mm (0.0008 to 0.0024 inch)
    Limit .... 0.10 mm (0.004 inch)
  Mainshaft fifth, countershaft first, second, third, reverse
    Standard .... 0.020 to 0.062 mm (0.0008 to 0.0024 inch)
    Limit .... 0.10 mm (0.004 inch)
Reverse idler gear-to-shaft clearance
  Standard .... 0.016 to 0.052 mm (0.0006 to 0.0020 inch)
  Limit .... 0.10 mm (0.004 inch)
Shaft diameters
  Mainshaft fourth gear surface
    Standard .... 19.959 to 19.980 mm (0.7858 to 0.7866 inch)
    Limit .... 19.93 mm (0.785 inch)
  Mainshaft fifth gear surface
    Standard .... 14.966 to 14.984 mm (0.5892 to 0.5899 inch)
    Limit .... 14.94 mm (0.588 inch)
  Reverse idler gear surface
    Standard .... 12.966 to 12.984 mm (0.5105 to 0.5112 inch)
    Limit .... 12.94 mm (0.509 inch)
Shaft-to-bushing clearances
  Mainshaft fourth, countershaft third
    Standard .... 0.020 to 0.062 mm (0.0008 to 0.0024 inch)
    Limit .... 0.10 mm (0.004 inch)
  Mainshaft fifth
    Standard .... 0.016 to 0.052 mm (0.0006 to 0.0020 inch)
    Limit .... 0.10 mm (0.004 inch)
Reverse idler shaft-to-gear clearance
  Standard .... 0.016 to 0.052 mm (0.0006 to 0.0020 inch)
  Limit .... 0.10 mm (0.004 inch)

Crankshaft

Connecting rod side clearance
  Standard .... 0.05 to 0.50 mm (0.002 to 0.020 inch)
  Limit .... 0.80 mm (0.031 inch)

Connecting rod big end radial clearance

Standard .... 0.004 to 0.012 mm (0.0002 to 0.0005 inch)

Limit .... 0.05 mm (0.002 inch)

Runout limit .... 0.06 mm (0.002 inch)

Torque specifications

Valve cover bolts (1997 through 2001 Recon)	Not specified
Rocker assembly bolts (2002-on Recon, all Rancher)	12 Nm (108 inch-lbs)
Cylinder head cap nuts	30 Nm (22 ft-lbs) (1)
Rocker assembly holder bolt	30 Nm (22 ft-lbs)
Rocker shaft retaining bolt (1997 through 2001 Recon)	7 Nm (61 inch-lbs)
Cam chain tensioner slider pivot bolt	12 Nm (108 inch-lbs) (3)
Front crankcase cover bolts	Not specified
Centrifugal clutch locknut	88 Nm (65 ft-lbs) (2) (5)
Change clutch spring bolts	12 Nm (108 inch-lbs)
Change clutch locknut	79 Nm (59 ft-lbs) (2)
Oil pump	
To-engine bolts	Not specified
Cover screw	Not specified
Shift pedal pinch bolt	18 Nm (156 inch-lbs)
External shift linkage return spring pin	22 Nm (12 ft-lbs) (3)
Gearshift cam plate bolt	16 Nm (144 inch-lbs) (3)
Stopper arm bolt	12 Nm (108 inch-lbs)
Gearshift arm-to-spindle bolt	25 Nm (18 ft-lbs) (4)
Crankcase bolts	Not specified
Cylinder studs to crankcase	6 Nm (52 inch-lbs)
Engine hanger nuts (both sides)	54 Nm (40 ft-lbs)

1. *Use engine oil on the threads and the seating surfaces of the nuts.*
2. *Use a new locknut with engine oil on the threads and the seating surface of the nut. Stake the locknut.*
3. *Use non-permanent thread locking agent on the threads.*
4. *Use a new lockwasher on 1997 through 2001 Recon and all TRX250EX models.*
5. *Left-hand threads.*

2.5 A compression gauge with a threaded fitting for the spark plug hole is preferred over the type that requires hand pressure to maintain the seal

1 General information

The engine/transmission unit is of the air-cooled, single-cylinder four-stroke design. The two valves are operated by pushrods and solid lifters. The lifters are operated by a camshaft mounted in the crankcase, which in turn is chain driven off the crankshaft. The engine/transmission assembly is constructed from aluminum alloy. The crankcase is divided vertically.

The crankshaft centerline is parallel to the centerline of the vehicle.

The crankcase incorporates a wet sump, pressure-fed lubrication system which uses a rotor-type oil pump, an oil filter and separate strainer screen and an oil temperature warning switch.

Power from the crankshaft is routed to the transmission via two clutches. The centrifugal clutch, which engages as engine speed is increased, connects the crankshaft to the change clutch, which is of the wet, multi-plate type. The change clutch transmits power to the transmission; it's engaged and disengaged automatically when the shift lever is moved from one gear position to another. The transmission has five forward gears and one reverse gear.

2 Cylinder compression - check

Refer to illustration 2.5
1   Among other things, poor engine performance may be caused by leaking valves, incorrect valve clearances, a leaking head gasket, or worn piston, rings and/or cylinder wall. A cylinder compression check will help pinpoint these conditions and can also indicate the presence of excessive carbon deposits in the cylinder head.
2   The only tools required are a compression gauge and a spark plug wrench. Depending on the outcome of the initial test, a squirt-type oil can may also be needed.
3   Start the engine and allow it to reach normal operating temperature, then remove the spark plug (see Chapter 1, if necessary). Work carefully - don't strip the spark plug hole threads and don't burn your hands.
4   Disable the ignition by disconnecting the primary (low tension) wires from the coil (see Chapter 4). Be sure to mark the locations of the wires before detaching them.
5   Install the compression gauge in the spark plug hole **(see illustration)**. Hold or block the throttle wide open.
6   Crank the engine over a minimum of four or five revolutions (or until the gauge reading stops increasing) and observe the initial movement of the compression gauge needle as well as the final total gauge reading. Compare the results to the value listed in this Chapter's Specifications.

7   If the compression built up quickly and evenly to the specified amount, you can assume the engine upper end is in reasonably good mechanical condition. Worn or sticking piston rings and worn cylinders will produce very little initial movement of the gauge needle, but compression will tend to build up gradually as the engine spins over. Valve and valve seat leakage, or head gasket leakage, is indicated by low initial compression which does not tend to build up.
8   To further confirm your findings, add a small amount of engine oil to the cylinder by inserting the nozzle of a squirt-type oil can through the spark plug hole. The oil will tend to seal the piston rings if they are leaking.
9   If the compression increases significantly after the addition of the oil, the piston rings and/or cylinder are definitely worn. If the compression does not increase, the pressure is leaking past the valves or the head gasket. Leakage past the valves may be due to insufficient valve clearances, burned, warped or cracked valves or valve seats or valves that are hanging up in the guides.
10   If compression readings are considerably higher than specified, the combustion chamber is probably coated with excessive carbon deposits. It is possible (but not very likely) for carbon deposits to raise the compression enough to compensate for the effects of leakage past rings or valves. Refer to Section 8, remove the cylinder head and carefully decarbonize the combustion chamber.

3 Operations possible with the engine in the frame

The components and assemblies listed below can be removed without having to remove the engine from the frame. If, however, a number of areas require attention at the same time, removal of the engine is recommended.

Valve cover
Cylinder head
Rocker arms and pushrods
Cylinder and piston
Gear selector mechanism external components
Recoil starter
Starter motor
Starter reduction gears
Starter clutch
Alternator rotor and stator
Clutches (centrifugal clutch and change clutch)
Cam chain tensioner
Camshaft
Oil pump

4 Operations requiring engine removal

It is necessary to remove the engine/transmission assembly from the frame and separate the crankcase halves to gain access to the following components:

Crankshaft and connecting rod
Transmission shafts
Shift drum and forks

5 Major engine repair - general note

1   It is not always easy to determine when or if an engine should be completely overhauled, as a number of factors must be considered.
2   High mileage is not necessarily an indication that an overhaul is needed, while low mileage, on the other hand, does not preclude the need for an overhaul. Frequency of servicing is probably the single most important consideration. An engine that has regular and frequent oil and filter changes, as well as other required maintenance, will most likely give many miles of reliable service. Conversely, a neglected engine, or one which has not been broken in properly, may require an overhaul very early in its life.

6.4 On 350 models and later 250 models, remove the screws and take off the engine side covers (350 right side cover shown; others similar)

6.5 On 350 models, disconnect the starter cable and two ground cables (arrows)

3   Exhaust smoke and excessive oil consumption are both indications that piston rings and/or valve guides are in need of attention. Make sure oil leaks are not responsible before deciding that the rings and guides are bad. Refer to Chapter 1 and perform a cylinder compression check to determine for certain the nature and extent of the work required.

4   If the engine is making obvious knocking or rumbling noises, the connecting rod and/or main bearings are probably at fault.

5   Loss of power, rough running, excessive valve train noise and high fuel consumption rates may also point to the need for an overhaul, especially if they are all present at the same time. If a complete tune-up does not remedy the situation, major mechanical work is the only solution.

6   An engine overhaul generally involves restoring the internal parts to the specifications of a new engine. During an overhaul the piston rings are replaced and the cylinder walls are bored and/or honed. If a rebore is done, then a new piston is also required. The crankshaft and connecting rod are permanently assembled, so if one of these components needs to be replaced both must be. Generally the valves are serviced as well, since they are usually in less than perfect condition at this point. While the engine is being overhauled, other components such as the carburetor and the starter motor can be rebuilt also. The end result should be a like-new engine that will give as many trouble-free miles as the original.

7   Before beginning the engine overhaul, read through all of the related procedures to familiarize yourself with the scope and requirements of the job. Overhauling an engine is not all that difficult, but it is time consuming. Plan on the vehicle being tied up for a minimum of two (2) weeks. Check on the availability of parts and make sure that any necessary special tools, equipment and supplies are obtained in advance.

8   Most work can be done with typical shop hand tools, although a number of precision measuring tools are required for inspecting parts to determine if they must be replaced. Often a dealer service department or repair shop will handle the inspection of parts and offer advice concerning reconditioning and replacement. As a general rule, time is the primary cost of an overhaul so it doesn't pay to install worn or substandard parts.

9   As a final note, to ensure maximum life and minimum trouble from a rebuilt engine, everything must be assembled with care in a spotlessly clean environment.

---

6   Engine - removal and installation

Note: *Engine removal and installation should be done with the aid of an assistant to avoid damage or injury that could occur if the engine is dropped. A hydraulic floor jack should be used to support and lower the engine if possible (they can be rented at low cost).*

Removal

Refer to illustrations 6.4, 6.5, 6.10 and 6.13a, 6.13b and 6.13c

1   Drain the engine oil and disconnect the spark plug wire (see Chapter 1).

2   Remove the fuel tank, carburetor, air cleaner housing and exhaust system (see Chapter 4). The choke and throttle cables can be left connected. Plug the carburetor intake openings with rags.

3   Remove the front fenders (see Chapter 7).

4   If you're working on a 350 model or a later 250 model, remove the bolts and collars and take the right and left side covers off the engine **(see illustration)**.

5   Disconnect the starter cable and two ground cables from the engine **(see illustration)** (see Chapter 8 if necessary).

6   Label and disconnect the following wires (refer to Chapters 4 or 8 for component location if necessary):

   Ignition pulse generator
   Alternator
   Oil temperature and gear position switches
   Speed sensor (if equipped)
   Shift control motor and angle sensor (electric shift models)

7   Disconnect the reverse cable (see Section 22).

8   Remove the shift pedal (if equipped) (see Section 24).

9   On 4wd models, remove the front driveshaft and loosen the clamps on the rear driveshaft boot that connects the swingarm to the engine (see Chapter 5).

10   Disconnect the crankcase breather tube **(see illustration)**.

6.10 Disconnect the crankcase breather tube (arrow) (350 shown; 250 similar)

6.13a On 350 models, remove the engine mount at the top

6.13b On all models, remove the mount at the lower right (350 shown) . . .

11   Disconnect the oil cooler hoses from the engine (see Section 18).

12   Support the engine securely from below. **Note:** *Use a jack (preferably a floor jack) that can be repositioned if necessary as the engine mounting bolts are removed.*

13   If you're working on a 350 model, remove the engine mounting bolts, nuts and brackets at the top **(see illustration)**. If you're working on a 250 model, unbolt the bracket that supports the air cleaner intake duct and upper engine bushing. On all models, remove the fasteners from the lower left and lower right **(see illustrations)**.

14   Have an assistant help you lift the engine. As you lift, pull the engine forward to disengage the rear driveshaft from the engine. The driveshaft is mounted inside the left side of the swingarm (refer to Chapter 5 if necessary). Remove the engine to the left side of the vehicle.

15   Slowly lower the engine to a suitable work surface.

Installation

16   Check the rubber engine supports for wear or damage and replace them if necessary before installing the engine. On 350 models, the upper hanger bracket is replaced as a unit if the rubber is damaged or deteriorated. On 250 models, the upper bushing (which fits inside a hole in the crossmember) can be replaced separately. On all models, the lower left and right bushings can be replaced separately.

17   Coat the driveshaft splines with moly-based grease. Install the engine from the left side of the vehicle and move it rearward to engage the driveshaft with the engine (refer to Chapter 5 if necessary).

18   Lift the engine to align the mounting bolt holes, then install the

upper bracket (350 models), and lower bolts, nuts and collars (all models), but don't tighten them yet. The wide sides of the dampers face the lower mounts.

19   Tighten the lower mounts (all models), then the upper mount 350 models) to the torques listed in this Chapter's Specifications.

20   The remainder of installation is the reverse of the removal steps, with the following additions:

a)   *Use new gaskets at all exhaust pipe connections.*

b)   *Use new O-rings at the oil cooler hose connections.*

c)   *Adjust the throttle cable and reverse selector cable following the procedures in Chapter 1.*

d)   *Fill the engine with oil and check the differential oil level, also following the procedures in Chapter 1. Run the engine and check for leaks.*

7   Engine disassembly and reassembly - general information

Refer to illustrations 7.2 and 7.3

1   Before disassembling the engine, clean the exterior with a degreaser and rinse it with water. A clean engine will make the job easier and prevent the possibility of getting dirt into the internal areas of the engine.

2   In addition to the precision measuring tools mentioned earlier, you will need a torque wrench, a valve spring compressor, oil gallery brushes **(see illustration)**, a piston ring removal and installation tool, a

6.13c . . . and lower left (350 shown)

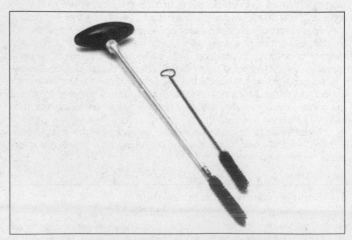

7.2 A selection of brushes is required for cleaning holes and passages in the engine components

7.3 An engine stand can be made from short lengths of lumber and lag bolts or nails

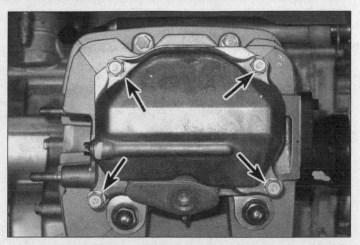

8.2 Remove the valve cover bolts (arrows) and lift off the cover (350 shown; 1997-2001 250 similar). . .

piston ring compressor and a clutch holder tool (described in Sections 16 and 18). Some new, clean engine oil of the correct grade and type, some engine assembly lube (or moly-based grease) and a tube of RTV (silicone) sealant will also be required.

3 An engine support stand make from short lengths of 2 x 4's bolted together will facilitate the disassembly and reassembly procedures **(see illustration)**. If you have an automotive-type engine stand, an adapter plate can be made from a piece of plate, some angle iron and some nuts and bolts.

4 When disassembling the engine, keep "mated" parts together (including gears, drum shifter pawls, etc.) that have been in contact with each other during engine operation. These "mated" parts must be reused or replaced as an assembly.

5 Engine/transmission disassembly should be done in the following general order with reference to the appropriate Sections.

Remove the valve cover, rocker arms and pushrods
Remove the cylinder head
Remove the cylinder
Remove the piston
Remove the clutches
Remove the oil pump
Remove the cam chain tensioner and camshaft
Remove the external shift mechanism
Remove the alternator rotor and starter clutch
Separate the crankcase halves
Remove the shift drum/forks
Remove the transmission shafts/gears
Remove the crankshaft and connecting rod

6 Reassembly is accomplished by reversing the general disassembly sequence.

---

8 Valve cover (1997 through 2001 Recon and all 350 models) - removal, inspection and installation

---

Note: *The valve cover can be removed with the engine in the frame. If the engine has been removed, ignore the steps which don't apply. The valve cover on 2002 and later Recon models, as well as all TRX250EX models, is an integral part of the rocker assembly and is removed as part of that procedure.*

Removal

Refer to illustrations 8.2 and 8.3

1 Remove the fuel tank and its heat shield (see Chapter 4).

2 Remove the valve cover bolts and lift off the cover **(see illustration)**. If it won't come easily, tap it from the side with a rubber mallet. Be careful not to damage the gasket surfaces of the cover or cylinder head by prying the cover loose.

3 Remove the valve cover gasket and O-ring **(see illustration)**.

4 Clean all traces of old gasket from the surfaces of the valve cover and cylinder head.

Installation

5 Install a new gasket and O-ring on the cylinder head. Install the valve cover bolts and tighten them evenly, in a criss-cross pattern. Tighten the bolts securely, but don't distort the valve cover or strip the threads.

---

9 Rocker arms, pushrods and cylinder head - removal, inspection and installation

---

Removal

1997 through 2001 Recon and all 350 models

Refer to illustrations 9.5, 9.6 and 9.9

1 Remove the carburetor, intake manifold, exhaust pipe and fuel tank mounting brackets (see Chapter 4). Remove the upper engine mounting bracket (see Section 6).

2 Remove the valve cover (see Section 7).

3 Place the piston at TDC on the compression stroke (see Valve clearances - adjustment in Chapter 1).

4 Disconnect the spark plug wire (see Chapter 1 if necessary).

5 Loosen the rocker assembly bolt, two cylinder head bolts and four cylinder head nuts in two or three stages, in a criss-cross pattern, then

8.3 . . . and remove the gasket and O-ring (arrow)

9.5 Remove the small bolts (upper two arrows) first, then loosen the nuts (lower four arrows) in a criss-cross pattern in two or three stages (350 shown; 1997 through 2001 250 similar)

9.6 Lift off the rocker assembly and locate the dowels (arrows) (350 shown; 1997 through 2001 250 similar)

remove the nuts and washers **(see illustration)**. **Note:** *Don't loosen the rocker holder bolt. It holds the rocker assembly together.*

6   Lift the rocker arm assembly off the cylinder head. Locate the dowels **(see illustration)**. They may have come off the rocker arm

assembly or remained in the cylinder head.

7   Label the pushrods so they can be reinstalled right-side up and in their original locations (exhaust or intake), then lift them from the engine **(see illustration 9.6)**.

8   Lift the cylinder head off the cylinder. If the head is stuck, tap around the side of the head with a rubber mallet to jar it loose, or use two wooden dowels inserted into the intake and exhaust ports to lever the head off. Don't attempt to pry the head off by inserting a screwdriver between the head and the cylinder block - you'll damage the sealing surfaces.

9   Cover the pushrod holes with a clean shop towel to prevent the entry of debris, and to make sure you don't drop the dowels down the pushrod passages. Once this is done, remove the head gasket and the two dowel pins from the cylinder **(see illustration)**. With the gasket out of the way, stuff a clean shop towel into the pushrod holes.

2002 and later Recon, all TRX250EX models
Refer to illustrations 9.13, 9.14 and 9.16

10   Remove the fuel tank and its heat shield (see Chapter 3).

11   On models without a recoil starter, unbolt the outer cover from the rear of the crankcase.

12   Place the crankshaft at top dead center (see the valve adjustment procedure in Chapter 1).

13   Loosen the rocker assembly bolts in three or four stages, in a criss-cross pattern, until they're all loose **(see illustration)**.

14   Lift off the rocker assembly, then remove the gasket and dowels **(see illustration)**.

9.9 Lift the cylinder head off and remove the gasket and dowels (arrows) (350 shown; 1997 through 2001 250 similar)

9.13 Loosen the rocker assembly bolts (arrows) in stages, in a criss-cross pattern

9.14 Lift off the rocker assembly, then remove the gasket and dowels (arrows)

9.16 **Loosen the cylinder head bolts and nuts (arrows) in a criss-cross pattern in several stages**

9.20a **Unscrew the rocker assembly bolt (arrow) . . .**

15   Remove the pushrods as described above.

16   Loosen the two cylinder head bolts and four nuts in a criss-cross pattern in several stages **(see illustration)**. When they're all loose, remove the bolts, nuts and washers.

17   Lift the cylinder head off the cylinder. If the head is stuck, tap around the side of the head with a rubber mallet to jar it loose, or use two wooden dowels inserted into the intake and exhaust ports to lever the head off. Don't attempt to pry the head off by inserting a screwdriver between the head and the cylinder block - you'll damage the sealing surfaces.

18   Cover the pushrod holes with a clean shop towel to prevent the entry of debris, and to make sure you don't drop the dowels down the pushrod passages. Once this is done, remove the head gasket and the two dowel pins from the cylinder **(see illustration 9.9)**. With the gasket out of the way, stuff a clean shop towel into the pushrod holes.

Inspection

Rocker assembly and pushrods

Refer to illustrations 9.20a, 9.20b, 9.21a, 9.21b, 9.26a and 9.26b

19   Label the rocker arms so they can be reinstalled in their original locations.

20   If you're working on a 350 or a 1997 through 2001 model Recon, unscrew the rocker assembly bolt. Slide out the rocker shaft and remove the rocker arms **(see illustrations)**.

21   If you're working on a 2002 or later Recon or a TXR250EX, unscrew the rocker shaft bolts and remove their sealing washers **(see**

9.20b **. . . slide out the shaft and remove the rocker arms (350 and 1997 through 2001 Recon)**

illustration). Remove the rocker shafts from the housing **(see illustration)**. If necessary, twist the shafts to free them using a flat-bladed screwdriver in the slot in the end of each shaft. You can also insert a screwdriver into the shaft bolt holes and pry the shafts out, but be careful not to damage the rocker housing.

9.21a **On 2002-on Recon and all TXR250EX models, unscrew the rocker shaft bolts and remove their sealing washers . . .**

9.21b **. . . then remove the rocker shafts and O-rings from the housing**

9.26a Remove the rocker shafts, noting the locations of the washers (arrows) . . .

9.26b . . . the washers are installed like this

22   Check the rocker arms for wear at the pushrod contact surfaces and at the tips of the valve adjusting screws. Try to twist the rocker arms from side-to-side on the shafts. If they're loose on the shafts or if there's visible wear, replace them.

23   Measure the outer diameter of each rocker shaft and the inner diameter of the rocker arms with a micrometer and compare the measurements to the values listed in this Chapter's Specifications. If rocker arm-to-shaft clearance is excessive, replace the rocker arm or shaft, whichever is worn.

24   Check the pushrod ends for wear or scoring. Roll the pushrods on a flat surface such as a piece of plate glass and check them for bending. Replace the pushrods if problems are visible. Don't disassemble them.

25   Coat the rocker shaft and rocker arm bores with moly-based grease. If you're working on a 2002 or later Recon or a TRX250EX, coat new rocker shaft O-rings with clean engine oil and install them on the shafts.

26   Install the rocker arms and shaft, making sure the rocker arms are returned to their original locations if you're reusing the old ones. On 2002 and later Recon and all TRX250EX models, make sure the spring washers are returned to their correct positions next to the rocker arms **(see illustrations)**. As you install the rocker shaft, align its hole with the bolt hole, using the screwdriver slot in the end of the shaft to twist it in its bore.

27   Install the rocker assembly bolt(s) and washer(s) and tighten to the torque listed in this Chapter's Specifications.

Cylinder head

28   Check the cylinder head gasket and the mating surfaces on the cylinder head and cylinder for leakage, which could indicate warpage. Refer to Section 10 and check the flatness of the cylinder head.

29   Clean all traces of old gasket material from the cylinder head and cylinder. Be careful not to let any of the gasket material fall into the crankcase, the cylinder bore or the bolt holes.

Installation

Refer to illustration 9.32

30   Install the two dowel pins, then lay the new gasket in place on the cylinder. Never reuse the old gasket and don't use any type of gasket sealant.

31   Carefully lower the cylinder head over the studs and dowels.

32   Install the pushrods. If you're reusing the old ones, return them to their original positions. Also be sure the ends of pushrods fit in the lifter pockets, not off to the side **(see illustration)**.

33   Make sure the rocker assembly dowels are in position **(see illustration 9.6)**, then install the rocker assembly over the studs

34   Install the washers and nuts on the studs.

35   Tighten the nuts and bolts in two or three stages to the torque listed in this Chapter's Specifications, working in a criss-cross pattern.

36   Install all parts removed for access.

37   Change the engine oil and adjust the valve clearances (see Chapter 1).

10   Valves/valve seats/valve guides - servicing

1   Because of the complex nature of this job and the special tools and equipment required, servicing of the valves, the valve seats and the valve guides (commonly known as a valve job) is best left to a professional.

2   The home mechanic can, however, remove and disassemble the head, do the initial cleaning and inspection, then reassemble and deliver the head to a dealer service department or properly equipped vehicle repair shop for the actual valve servicing. Refer to Section 10 for those procedures.

3   The dealer service department will remove the valves and springs, recondition or replace the valves and valve seats, replace the valve guides, check and replace the valve springs, spring retainers and keepers (as necessary), replace the valve seals with new ones and reassemble the valve components.

4   After the valve job has been performed, the head will be in like-new condition. When the head is returned, be sure to clean it again very thoroughly before installation on the engine to remove any metal particles or abrasive grit that may still be present from the valve service operations. Use compressed air, if available, to blow out all the holes and passages.

9.32 Be sure to install the pushrod ends in the pockets of the lifters, not off to the side

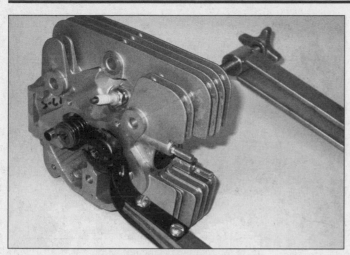

11.7a Compress the spring, remove the keepers and release the compressor

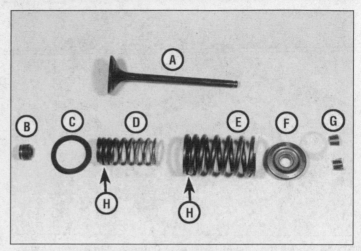

11.7b Valve and related components

A	Valve stem	E	Outer valve spring
B	Oil seal	F	Valve spring retainer
C	Spring seat	G	Keepers
D	Inner valve spring	H	Tightly wound coils

11 Cylinder head and valves - disassembly, inspection and reassembly

1   As mentioned in the previous Section, valve servicing and valve guide replacement should be left to a dealer service department or vehicle repair shop. However, disassembly, cleaning and inspection of the valves and related components can be done (if the necessary special tools are available) by the home mechanic. This way no expense is incurred if the inspection reveals that service work is not required at this time.

2   To properly disassemble the valve components without the risk of damaging them, a valve spring compressor is absolutely necessary. If the special tool is not available, have a dealer service department or vehicle repair shop handle the entire process of disassembly, inspection, service or repair (if required) and reassembly of the valves.

Disassembly

Refer to illustrations 11.7a, 11.7b and 11.7c

3   Remove the carburetor intake tube from the cylinder head (see Chapter 3).

4   Before the valves are removed, scrape away any traces of gasket material from the head gasket sealing surface. Work slowly and do not nick or gouge the soft aluminum of the head. Gasket removing solvents, which work very well, are available at most vehicle shops and auto parts stores.

5   Carefully scrape all carbon deposits out of the combustion chamber area. A hand held wire brush or a piece of fine emery cloth can be used once most of the deposits have been scraped away. Do not use a wire brush mounted in a drill motor, or one with extremely stiff bristles, as the head material is soft and may be eroded away or scratched by the wire brush.

6   Before proceeding, arrange to label and store the valves along with their related components so they can be kept separate and reinstalled in the same valve guides they are removed from (again, plastic bags work well for this).

7   Compress the valve spring(s) on the first valve with a spring compressor, then remove the keepers and the retainer from the valve assembly **(see illustration)**. Do not compress the spring(s) any more than is absolutely necessary. Carefully release the valve spring compressor and remove the spring(s), spring seat and valve from the head **(see illustration)**. If the valve binds in the guide (won't pull through), push it back into the head and deburr the area around the keeper groove with a very fine file or whetstone **(see illustration)**.

8   Repeat the procedure for the remaining valves. Remember to keep the parts for each valve together so they can be reinstalled in the same location.

9   Once the valves have been removed and labeled, pull off the valve stem seals with pliers and discard them (the old seals should never be reused).

10   Next, clean the cylinder head with solvent and dry it thoroughly. Compressed air will speed the drying process and ensure that all holes and recessed areas are clean.

11   Clean all of the valve springs, keepers, retainers and spring seats with solvent and dry them thoroughly. Do the parts from one valve at a time so that no mixing of parts between valves occurs.

12   Scrape off any deposits that may have formed on the valve, then use a motorized wire brush to remove deposits from the valve heads and stems. Again, make sure the valves do not get mixed up.

Inspection

Refer to illustrations 11.14, 11.15, 11.16, 11.17, 11.18, 11.19a and 11.19b

13   Inspect the head very carefully for cracks and other damage. If cracks are found, a new head will be required. Check the cam bearing surfaces for wear and evidence of seizure. Check the camshaft for wear as well (see Section 20).

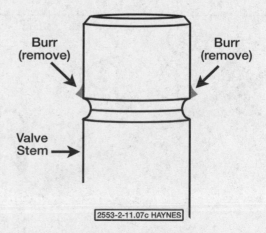

11.7c Check the area around the keeper groove for burrs and remove any that you find

11.14 Check the gasket surface for flatness with a straightedge and feeler gauge in the directions shown

11.15 Measuring valve seat width

14   Using a precision straightedge and a feeler gauge, check the head gasket mating surface for warpage. Lay the straightedge lengthwise, across the head and diagonally (corner-to-corner), intersecting the head bolt holes, and try to slip a feeler gauge under it, on either side of each combustion chamber **(see illustration)**. The feeler gauge thickness should be the same as the cylinder head warpage limit listed in this Chapter's Specifications. If the feeler gauge can be inserted between the head and the straightedge, the head is warped and must either be machined or, if warpage is excessive, replaced with a new one.

15   Examine the valve seats in each of the combustion chambers. If they are pitted, cracked or burned, the head will require valve service that is beyond the scope of the home mechanic. Measure the valve seat width **(see illustration)** and compare it to this Chapter's Specifications. If it is not within the specified range, or if it varies around its circumference, valve service work is required.

16   Clean the valve guides to remove any carbon buildup, then measure the inside diameters of the guides (at both ends and the center of the guide) with a small hole gauge and a micrometer **(see illustration)**. Record the measurements for future reference. The guides are measured at the ends and at the center to determine if they are worn in a bell-mouth pattern (more wear at the ends). If they are, guide replacement is an absolute must.

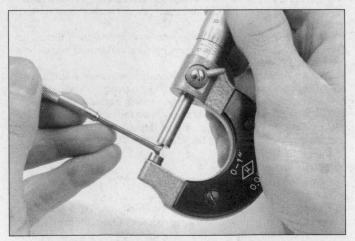

11.16 Measure the valve guide inside diameter with a hole gauge, then measure the gauge with a micrometer

17   Carefully inspect each valve face for cracks, pits and burned spots. Check the valve stem and the keeper groove area for cracks **(see illustration)**. Rotate the valve and check for any obvious indication that it is bent. Check the end of the stem for pitting and excessive wear and make sure the bevel is the specified width. The presence of any of the above conditions indicates the need for valve servicing.

18   Measure the valve stem diameter **(see illustration)**. If the diameter is less than listed in this Chapter's Specifications, the valves will have to be replaced with new ones. Also check the valve stem for bending. Set the valve in a V-block with a dial indicator touching the middle of the stem. Rotate the valve and look for a reading on the gauge (which

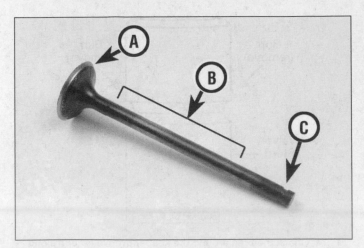

11.17 Check the valve face (A), stem (B) and keeper groove (C) for wear and damage

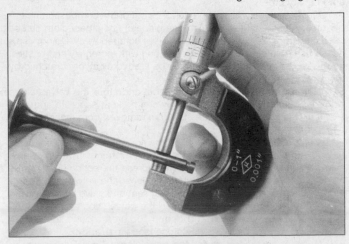

11.18 Measuring valve stem diameter

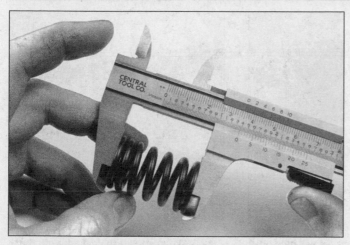

11.19a Measuring the free length of the valve springs

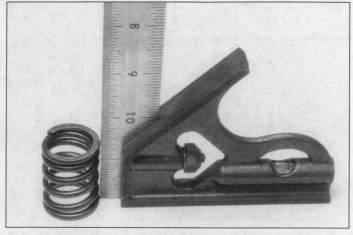

11.19b Checking the valve springs for squareness

indicates a bent stem). If the stem is bent, replace the valve.

19   Check the end of each valve spring for wear and pitting. Measure the free length **(see illustration)** and compare it to this Chapter's Specifications. Any springs that are shorter than specified have sagged and should not be reused. Stand the spring on a flat surface and check it for squareness **(see illustration)**.

20   Check the spring retainers and keepers for obvious wear and cracks. Any questionable parts should not be reused, as extensive damage will occur in the event of failure during engine operation.

21   If the inspection indicates that no service work is required, the valve components can be reinstalled in the head.

Reassembly

Refer to illustrations 11.23, 11.24a, 11.24b, 11.26 and 11.27

22   If the valve seats have been ground, the valves and seats should be lapped before installing the valves in the head to ensure a positive seal between the valves and seats. This procedure requires coarse and fine valve lapping compound (available at auto parts stores) and a valve lapping tool. If a lapping tool is not available, a piece of rubber or plastic hose can be slipped over the valve stem (after the valve has been installed in the guide) and used to turn the valve.

23   Apply a small amount of coarse lapping compound to the valve face **(see illustration)**, then slip the valve into the guide. **Note:** *Make sure the valve is installed in the correct guide and be careful not to get any lapping compound on the valve stem.*

24   Attach the lapping tool (or hose) to the valve and rotate the tool between the palms of your hands. Use a back-and-forth motion rather

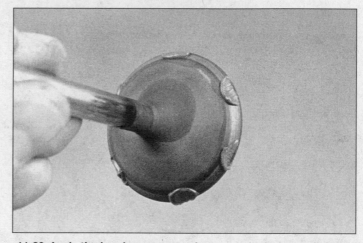

11.23 Apply the lapping compound very sparingly, in small dabs, to the valve face only

than a circular motion. Lift the valve off the seat and turn it at regular intervals to distribute the lapping compound properly. Continue the lapping procedure until the valve face and seat contact area is of uniform width and unbroken around the entire circumference of the valve face and seat **(see illustrations)**. Once this is accomplished, lap the valves again with fine lapping compound.

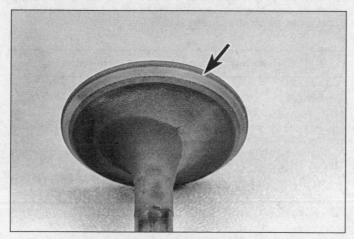

11.24a After lapping, the valve face should exhibit a uniform, unbroken contact pattern (arrow) . . .

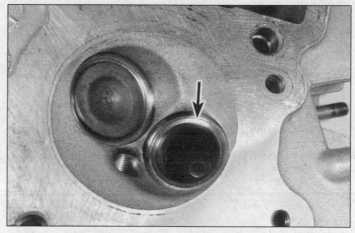

11.24b . . . and the seat (arrow) should be the specified width with a smooth, unbroken appearance

11.26 Push the oil seal onto the valve guide (arrow)

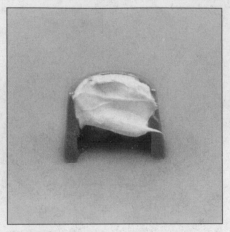

11.27 A small dab of grease will help hold the keepers in place on the valve while the spring compressor is released

12.3 When you lift the cylinder off, the dowels (arrows) may come off with the cylinder or stay in the crankcase

12.4a Here are the 350 dowel locations (right arrows) and sealant locations (left arrows)

12.4b Here are the 250 dowel locations (lower arrows) and sealant locations (upper arrows)

25   Carefully remove the valve from the guide and wipe off all traces of lapping compound. Use solvent to clean the valve and wipe the seat area thoroughly with a solvent soaked cloth. Repeat the procedure for the remaining valves.

26   Lay the spring seat in place in the cylinder head, then install new valve stem seals on both of the guides **(see illustration)**. Use an appropriate size deep socket to push the seals into place until they are properly seated. Don't twist or cock them, or they will not seal properly against the valve stems. Also, don't remove them again or they will be damaged.

27   Coat the valve stems with assembly lube or moly-based grease, then install one of them into its guide. Next, install the spring seat, springs and retainers, compress the springs and install the keepers. **Note:** *Install the springs with the tightly wound coils at the bottom (next to the spring seat).* When compressing the springs with the valve spring compressor, depress them only as far as is absolutely necessary to slip the keepers into place. Apply a small amount of grease to the keepers **(see illustration)** to help hold them in place as the pressure is released from the springs. Make certain that the keepers are securely locked in their retaining grooves.

28   Support the cylinder head on blocks so the valves can't contact the workbench top, then very gently tap each of the valve stems with a soft-faced hammer. This will help seat the keepers in their grooves.

29   Once all of the valves have been installed in the head, check for

proper valve sealing by pouring a small amount of solvent into each of the valve ports. If the solvent leaks past the valve(s) into the combustion chamber area, disassemble the valve(s) and repeat the lapping procedure, then reinstall the valve(s) and repeat the check. Repeat the procedure until a satisfactory seal is obtained.

12 Cylinder - removal, inspection and installation

Removal

Refer to illustrations 12.3, 12.4a, 12.4b and 12.5

1   Remove the valve cover, rocker arms, pushrods and cylinder head (see Sections 7 and 8).

2   Make sure the crankshaft is positioned at Top Dead Center (TDC).

3   Lift the cylinder straight up to remove it **(see illustration)**. If it's stuck, tap around its perimeter with a soft-faced hammer (but don't tap on the cooling fins or they may break). Don't attempt to pry between the cylinder and the crankcase, as you'll ruin the sealing surfaces.

4   Locate the dowel pins (they may have come off with the cylinder or still be in the crankcase) **(see illustrations)**. Be careful not to let these drop into the engine.

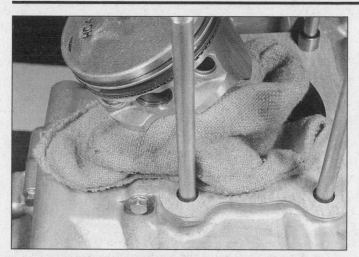

12.5 Pack clean rags into the crankcase opening to keep out debris

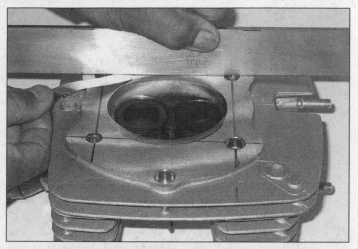

12.6 Check the cylinder top surface for warpage in the directions shown

5   Stuff rags around the piston **(see illustration)** and remove the gasket and all traces of old gasket material from the surfaces of the cylinder and the crankcase.

Inspection

Refer to illustration 12.6

Caution: *Don't attempt to separate the liner from the cylinder.*

6   Check the top surface of the cylinder for warpage, using the same method as for the cylinder head (see Section 10). Measure along the sides and diagonally across the stud holes **(see illustration)**.

7   Check the cylinder walls carefully for scratches and score marks.

8   Using the appropriate precision measuring tools, check the cylinder's diameter at the top, center and bottom of the cylinder bore, parallel to the crankshaft axis. Next, measure the cylinder's diameter at the same three locations across the crankshaft axis. Compare the results to this Chapter's Specifications. If the cylinder walls are tapered, out-of-round, worn beyond the specified limits, or badly scuffed or scored, have the cylinder rebored and honed by a dealer service department or an ATV repair shop. If a rebore is done, oversize pistons and rings will be required as well.

9   As an alternative, if the precision measuring tools are not available, a dealer service department or repair shop will make the measurements and offer advice concerning servicing of the cylinder.

10   If it's in reasonably good condition and not worn to the outside of the limits, and if the piston-to-cylinder clearance can be maintained properly, then the cylinder does not have to be rebored; honing is all that is necessary.

11   To perform the honing operation you will need the proper size flexible hone with fine stones as shown in Maintenance techniques, tools and working facilities at the front of this book, or a "bottle brush" type hone, plenty of light oil or honing oil, some shop towels and an electric drill motor. Hold the cylinder block in a vise (cushioned with soft jaws or wood blocks) when performing the honing operation. Mount the hone in the drill motor, compress the stones and slip the hone into the cylinder. Lubricate the cylinder thoroughly, turn on the drill and move the hone up and down in the cylinder at a pace which will produce a fine crosshatch pattern on the cylinder wall with the crosshatch lines intersecting at approximately a 60-degree angle. Be sure to use plenty of lubricant and do not take off any more material than is absolutely necessary to produce the desired effect. Do not withdraw the hone from the cylinder while it is running. Instead, shut off the drill and continue moving the hone up and down in the cylinder until it comes to a complete stop, then compress the stones and withdraw the hone. Wipe the oil out of the cylinder and repeat the procedure on the remaining cylinder. Remember, do not remove too much material from the cylinder wall. If you do not have the tools, or do not desire to perform the honing operation, a dealer service department or vehicle repair shop will generally do it for a reasonable fee.

12   Next, the cylinder must be thoroughly washed with warm soapy water to remove all traces of the abrasive grit produced during the honing operation. Be sure to run a brush through the bolt holes and flush them with running water. After rinsing, dry the cylinder thoroughly and apply a coat of light, rust-preventative oil to all machined surfaces.

Installation

Refer to illustration 12.16

13   Lubricate the cylinder bore with plenty of clean engine oil. Apply a thin film of moly-based grease to the piston skirt.

14   Thoroughly clean the cylinder mating surface on the crankcase. Apply a small dab of silicone sealant to each of the three points where the crankcase seam meets the surface **(see illustration 12.4a or 12.4b)**. Install the dowel pins, then lower a new cylinder base gasket over them.

15   Attach a piston ring compressor to the piston and compress the piston rings. A large hose clamp can be used instead - just make sure it doesn't scratch the piston, and don't tighten it too much.

16   Install the cylinder over the studs and carefully lower it down until the piston crown fits into the cylinder liner **(see illustration)**. Push down on the cylinder, making sure the piston doesn't get cocked sideways, until the bottom of the cylinder liner slides down past the piston rings. A wood or plastic hammer handle can be used to gently tap the cylinder down, but don't use too much force or the piston will be damaged.

12.16 If you're very careful, the cylinder can be installed over the rings without a ring compressor, but a compressor is recommended

13.3a The IN mark on top of the piston faces the intake (rear) side of the engine

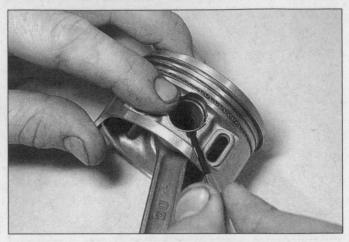

13.3b Wear eye protection and pry the circlip out of its groove with a pointed tool

17   Remove the piston ring compressor or hose clamp, being careful not to scratch the piston.
18   The remainder of installation is the reverse of the removal steps.

13 Piston - removal, inspection and installation

1   The piston is attached to the connecting rod with a piston pin that is a slip fit in the piston and rod.
2   Before removing the piston from the rod, stuff a clean shop towel into the crankcase hole, around the connecting rod. This will prevent the circlips from falling into the crankcase if they are inadvertently dropped.

Removal

Refer to illustrations 13.3a, 13.3b, 13.4a and 13.4b
3   The piston should have an IN mark on its crown that goes toward the intake (rear) side of the engine **(see illustration)**. If this mark is not visible due to carbon buildup, scribe an arrow into the piston crown before removal. Support the piston and pry the circlip out with a pointed tool **(see illustration)**.
4   Push the piston pin out from the opposite end to free the piston from the rod **(see illustration)**. You may have to deburr the area around the groove to enable the pin to slide out (use a triangular file for this procedure). If the pin won't come out, you can fabricate a piston pin

removal tool from a long bolt, a nut, a piece of tubing and washers **(see illustration)**.

Inspection

Refer to illustrations 13.6, 13.13, 13.14 and 13.16
5   Before the inspection process can be carried out, the pistons must be cleaned and the old piston rings removed.
6   Using a piston ring removal and installation tool, carefully remove the rings from the pistons **(see illustration)**. Do not nick or gouge the pistons in the process.
7   Scrape all traces of carbon from the tops of the pistons. A hand-held wire brush or a piece of fine emery cloth can be used once the majority of the deposits have been scraped away. Do not, under any circumstances, use a wire brush mounted in a drill motor to remove deposits from the pistons; the piston material is soft and will be eroded away by the wire brush.
8   Use a piston ring groove cleaning tool to remove any carbon deposits from the ring grooves. If a tool is not available, a piece broken off the old ring will do the job. Be very careful to remove only the carbon deposits. Do not remove any metal and do not nick or gouge the sides of the ring grooves.
9   Once the deposits have been removed, clean the pistons with solvent and dry them thoroughly. Make sure the oil return holes below the oil ring grooves are clear.
10   If the pistons are not damaged or worn excessively and if the cyl-

13.4a Push the piston pin partway out, then pull it the rest of the way

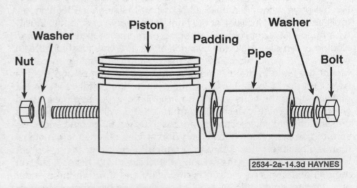

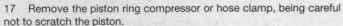

13.4b The piston pin should come out with hand pressure - if it doesn't, this removal tool can be fabricated from readily available parts

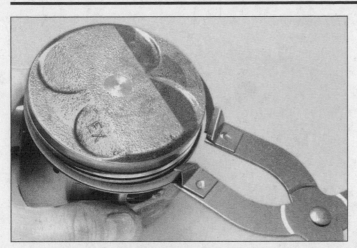

13.6 Remove the piston rings with a ring removal and installation tool

13.13 Measure the piston ring-to-groove clearance with a feeler gauge

inders are not rebored, new pistons will not be necessary. Normal piston wear appears as even, vertical wear on the thrust surfaces of the piston and slight looseness of the top ring in its groove. New piston rings, on the other hand, should always be used when an engine is rebuilt.

11   Carefully inspect each piston for cracks around the skirt, at the pin bosses and at the ring lands.

12   Look for scoring and scuffing on the thrust faces of the skirt, holes in the piston crown and burned areas at the edge of the crown. If the skirt is scored or scuffed, the engine may have been suffering from overheating and/or abnormal combustion, which caused excessively high operating temperatures. The oil pump should be checked thoroughly. A hole in the piston crown, an extreme to be sure, is an indication that abnormal combustion (pre-ignition) was occurring. Burned areas at the edge of the piston crown are usually evidence of spark knock (detonation). If any of the above problems exist, the causes must be corrected or the damage will occur again.

13   Measure the piston ring-to-groove clearance (side clearance) by laying a new piston ring in the ring groove and slipping a feeler gauge in beside it **(see illustration)**. Check the clearance at three or four locations around the groove. Be sure to use the correct ring for each groove; they are different. If the clearance is greater then specified, new pistons will have to be used when the engine is reassembled.

14   Check the piston-to-bore clearance by measuring the bore (see Section 11) and the piston diameter **(see illustration)**. Measure the piston across the skirt on the thrust faces at a 90-degree angle to the piston pin, at the specified distance up from the bottom of the skirt. Subtract the piston diameter from the bore diameter to obtain the clearance. If it is greater than specified, the cylinder will have to be rebored and a new oversized piston and rings installed. If the appropriate precision measuring tools are not available, the piston-to-cylinder clearance can be obtained, though not quite as accurately, using feeler gauge stock. Feeler gauge stock comes in 12-inch lengths and various thicknesses and is generally available at auto parts stores. To check the clearance, slip a piece of feeler gauge stock of the same thickness as the specified piston clearance into the cylinder along with appropriate piston. The cylinder should be upside down and the piston must be positioned exactly as it normally would be. Place the feeler gauge between the piston and cylinder on one of the thrust faces (90-degrees to the piston pin bore). The piston should slip through the cylinder (with the feeler gauge in place) with moderate pressure. If it falls through, or slides through easily, the clearance is excessive and a new piston will be required. If the piston binds at the lower end of the cylinder and is loose toward the top, the cylinder is tapered, and if tight spots are encountered as the piston/feeler gauge is rotated in the cylinder, the cylinder is out-of-round. Be sure to have the cylinder and piston checked by a dealer service department or a repair shop to confirm your findings before purchasing new parts.

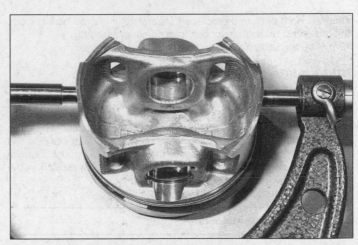

13.14 Measure the piston diameter with a micrometer

15   Apply clean engine oil to the pin, insert it partway into the piston and check for freeplay by rocking the pin back-and-forth. If the pin is loose, a new piston and possibly new pin must be installed.

16   Repeat Step 15, this time inserting the piston pin into the connecting rod **(see illustration)**. If the pin is loose, measure the pin diameter and the pin bore in the rod (or have this done by a dealer or repair

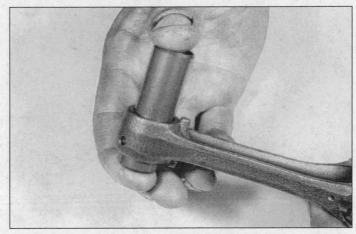

13.16 Slip the piston pin into the rod and try to rock it back-and-forth to check for looseness

13.18 Make sure both piston pin circlips are securely seated in the piston grooves

14.2 Check the piston ring end gap with a feeler gauge at the bottom of the cylinder

shop). A worn pin can be replaced separately; if the rod bore is worn, the rod and crankshaft must be replaced as an assembly.

17   Refer to Section 13 and install the rings on the pistons.

Installation

Refer to illustration 13.18

18   Install the piston with its IN mark toward the intake side (rear) of the engine. Lubricate the pin and the rod bore with moly-based grease. Install a new circlips in the groove in one side of the piston (don't reuse the old circlips). Push the pin into position from the opposite side and install another new circlip. Compress the circlips only enough for them to fit in the piston. Make sure the clips are properly seated in the grooves **(see illustration)**.

14 Piston rings - installation

Refer to illustrations 14.2, 14.4, 14.7a, 14.7b, and 14.7c

1   Before installing the new piston rings, the ring end gaps must be checked.

2   Insert the top (No. 1) ring into the bottom of the first cylinder and square it up with the cylinder walls by pushing it in with the top of the piston. The ring should be about one-half inch above the bottom edge of the cylinder. To measure the end gap, slip a feeler gauge between the ends of the ring **(see illustration)** and compare the measurement to the Specifications.

3   If the gap is larger or smaller than specified, double check to make sure that you have the correct rings before proceeding.

4   If the gap is too small, it must be enlarged or the ring ends may come in contact with each other during engine operation, which can cause serious damage. The end gap can be increased by filing the ring ends very carefully with a fine file **(see illustration)**. When performing this operation, file only from the outside in.

5   Repeat the procedure for the second compression ring (ring gap is not specified for the oil ring rails or spacer).

6   Once the ring end gaps have been checked/corrected, the rings can be installed on the piston.

7   The oil control ring (lowest on the piston) is installed first. It is composed of three separate components. Slip the spacer into the groove, then install the upper side rail **(see illustrations)**. Do not use a piston ring installation tool on the oil ring side rails as they may be damaged. Instead, place one end of the side rail into the groove between the spacer expander and the ring land. Hold it firmly in place and slide a finger around the piston while pushing the rail into the groove (taking care not to cut your fingers on the sharp edges). Next, install the lower side rail in the same manner.

8   After the three oil ring components have been installed, check to make sure that both the upper and lower side rails can be turned smoothly in the ring groove.

9   Install the no. 2 (middle) ring next. It can be readily distinguished from the top ring by its cross-section shape **(see illustration 14.7c)**. Do not mix the top and middle rings.

14.4 If the end gap is too small, clamp a file in a vise and file the ring ends (from the outside in only) to enlarge the gap slightly

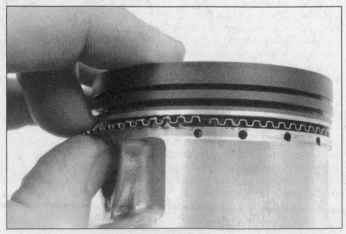

14.7a Installing the oil ring expander - make sure the ends don't overlap

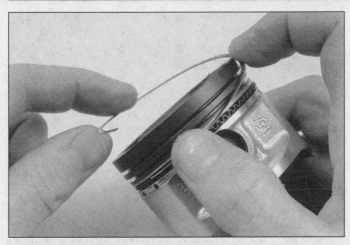

14.7b Installing an oil ring side rail - don't use a ring installation tool to do this

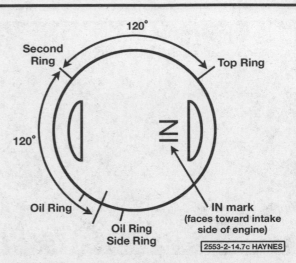

14.7c Ring details

10   To avoid breaking the ring, use a piston ring installation tool and make sure that the identification mark is facing up **(see illustration 14.7c)**. Fit the ring into the middle groove on the piston. Do not expand the ring any more than is necessary to slide it into place.

11   Finally, install the no. 1 (top) ring in the same manner. Make sure the identifying mark is facing up. Be very careful not to confuse the top and second rings. Besides the different profiles, the top ring is narrower than the second ring.

12   Once the rings have been properly installed, stagger the end gaps, including those of the oil ring side rails **(see illustration 14.7c)**.

15   Valve lifters - removal, inspection and installation

Removal

Refer to illustration 15.2

1   Remove the valve cover, rocker arms and pushrods, cylinder head and cylinder (see Sections 8 and 9).

2   Label the lifters so they can be returned to their correct bores **(see illustration)**. Lift the lifters out.

Inspection

3   Check the lifters for wear, galling, scoring or the bluish tint that indicates overheating. Check the camshaft contact surfaces, the sides and the pockets that contain the pushrod ends. If any problems can be seen, replace the lifters.

4   If there's any doubt about lifter wear, measure lifter diameter with a micrometer. Measure the internal diameter of the lifter bores, using a hole gauge and micrometer. If the lifter measurements are not within the range listed in this Chapter's Specifications, replace the lifters. If the bores are worn, replace the crankcase.

Installation

5   Lubricate the lifters on the camshaft contact surfaces, sides and pushrod pockets with molybdenum disulfide grease. Install the lifters in their bores.

6   The remainder of installation is the reverse of the removal steps.

16   Centrifugal clutch - removal, inspection and installation

Removal

Refer to illustrations 16.3, 16.5a, 16.5b, 16.5c, 16.5d, 16.6, 16.7a, 16.7b and 16.8

1   Drain the engine oil (see Chapter 1). Leave the dipstick out of the engine.

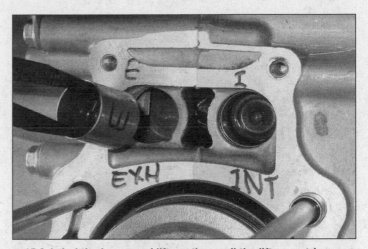

15.2 Label the bores and lifters, then pull the lifters out (use a magnet if necessary)

2   Disconnect the oil cooler hoses (see Section 19).

3   Remove the left engine cover and its bracket **(see illustration)**.

4   If you're working on an electric shift 4wd Rancher or 2001 and later electric shift Recon, remove the front driveshaft (Rancher only), shift control motor and shift control motor gears (see Chapters 5 and 8).

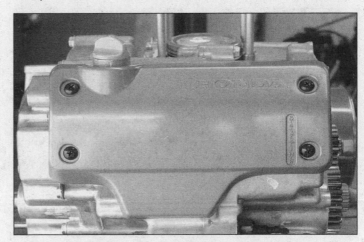

16.3 If you're working on a 350, remove the left engine cover and bracket

16.5a Remove the cover bolts (arrows) and take off the cover (here's a 350) . . .

16.5b . . . and here's a 250

5   Remove the front crankcase cover bolts **(see illustrations)**. Separate the cover from the engine, taking care not to damage the gasket surfaces. Remove the cover gasket and O-ring and locate the dowel pins **(see illustrations)**. The dowels may have come off with the cover or stayed in the engine. If you're working on an electric shift 4wd model, also locate the joint collars and O-rings and the joint shaft and washer. As with the dowels, they may have come off with the cover or stayed in the engine.

6   Grind or file away the staked portion of the clutch nut, taking care

16.5c On 350 models, locate the dowels (A) and oil pump collars and O-rings (B) . . .

16.5d . . . and here are the 250 dowels (arrows) - if the thrust washer falls off the shift mechanism, be sure to reinstall it

16.6 Grind or file away the staked portion of the locknut lip (arrow)

16.7a Hold the clutch with a tool like this one and unscrew the nut (250 models have left-hand threads)

16.7b Remove the washer; its OUT SIDE mark faces (if equipped) away from the engine

16.8 If the clutch won't come off easily, remove it with a puller like this one

16.9 While holding the drum, it should be possible to turn the weight assembly only in the direction of the arrow; if it turns both ways or neither way, remove it and inspect the one-way clutch

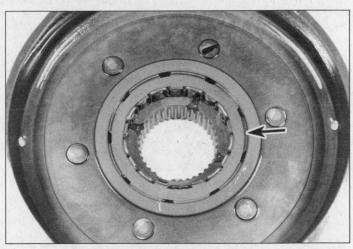

16.11 Lift the one-way clutch (arrow) out of the drum and check it for wear and damage

not to get metal particles in the engine **(see illustration)**.

7   Hold the clutch with a removal tool so it won't turn **(see illustrations)**. Remove the nut and washer. On 250 models, turn the nut clockwise to loosen it.

8   Pull the centrifugal clutch weight assembly and drum off the crankshaft. If it won't come easily, use a special tool (Honda part no. 07933-HB3000A or equivalent) **(see illustration)**.

Inspection

Refer to illustrations 16.9, 16.11, 16.12, 16.13a, 16.13b, 16.13c, 16.13d, 16.14, 16.21, 16.22 and 16.23

9   Hold the drum in one hand and try to rotate the weight assembly with the other hand **(see illustration)**. It should rotate counterclockwise only. If it rotates both ways or neither way, disassemble the centrifugal clutch for further inspection.

10   Lift the weight assembly out of the drum.

11   Lift the one-way clutch out of the drum **(see illustration)**. Check the one-way clutch rollers for signs of wear or scoring. The rotors should be unmarked with no signs of wear such as pitting or flat spots. Replace the one-way clutch if it's worn.

12   Measure the thickness of the lining material on the weights at the three thickest points **(see illustration)**. If it's thinner than the minimum listed in this Chapter's Specifications, replace the weights as a set.

16.12 Measure the thickness of the friction material at the three thick points on each weight (arrows)

16.13a Pry the snap-rings loose from the posts

16.13b Lift off the outer plate . . .

16.13c . . . the clutch spring (arrow) . . .

13   Pry the clips off the drive plate posts **(see illustration)**. Lift off the outer plate, clutch spring and inner plate **(see illustrations)**.
14   Check the springs for breakage and the weights for wear or dam-

age and replace them as necessary. If the weights need to be removed from the drive plate, unhook the springs **(see illustration)** and lift the weights off the posts.
15   Measure the inside diameter of the drum and replace if it it's greater than the limit listed in this Chapter's Specifications.
16   Measure the free length of the clutch weight springs and replace them if they've stretched to longer than the limit listed in this Chapter's Specifications.
17   Check the drive plate posts for wear or damage and replace the drive plate if problems are found.
18   Place the clutch spring on a flat surface (such as a piece of glass) with its concave side down. Measure the height of the spring center from the surface with a vernier caliper. If the spring has been flattened to less than the limit listed in this Chapter's Specifications, replace it.
19   Place the weights in the drive plate with their OUT SIDE marks facing upward **(see illustration 16.14)**. Install the springs, noting how their ends are located.
20   Place the inner plate on the weights with its lip upward **(see illustration 16.13d)**.
21   Place the clutch spring on the inner plate with its concave side toward the plate **(see illustration)**.
22   Place the outer plate on the clutch spring with its dimples upward **(see illustration)**.
23   Place a pair of C-clamps or similar clamps on the outer plate and tighten them until the circlip grooves in the posts are exposed, then

16.13d . . . and the inner plate; the lip on the inner plate (arrow) faces away from the weight assembly

16.14 Remove the springs and slip the weights off the drive plate posts; the OUT SIDE mark on each weight faces away from the drive plate

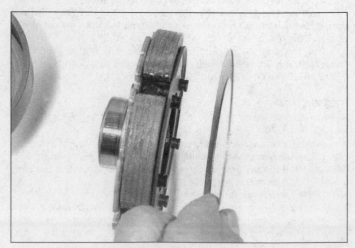

16.21 Install the clutch spring with its concave side toward the weight assembly

16.22 Install the outer plate with its locating pins (arrows) facing away from the weight assembly

16.23 Place a pair of clamps on the assembly and compress it so the post grooves are exposed, then squeeze the snap-rings into place with a pair of pliers

squeeze the clips onto the posts with pliers **(see illustration)**.

24   Install the one-way clutch in the drum with its OUT SIDE marking facing out **(see illustration 16.13d)**. Lubricate the one-way clutch with engine oil.

25   Install the weight assembly in the drum **(see illustration 16.9)**.

26   Refer to Section 26 and check the bearings and oil seals in the front cover for wear or damage.

Installation

27   Slip the centrifugal clutch onto the crankshaft.

28   Engage the drum splines with the splines on the crankshaft. Rotate the weight assembly slightly and align the drive plate splines with the primary gear, then push the centrifugal clutch all the way on.

29   Install the washer **(see illustration 16.7b)**.

30   Oil the threads of a new locknut, then install the locknut. Hold the clutch with a holding tool and tighten the nut to the torque listed in this Chapter's Specifications. 250 models have left-hand threads.

31   Stake the lip of the locknut with a hammer and punch **(see illustration 16.6)**.

32   Make sure the dowels are in position and install a new gasket **(see illustration 16.5b)**.

33   Thread the cover bolts into their holes **(see illustration 16.5a)**. Tighten the cover bolts in two or three stages, in a criss-cross pattern,

to the torque listed in this Chapter's Specifications.

34   The remainder of installation is the reverse of the removal steps.

35   Refill the engine oil (see Chapter 1).

17   Change clutch and release mechanism - removal, inspection and installation

Release mechanism (all except 2006 and later TRX250EX)

Removal

Refer to illustrations 17.2a, 17.2b and 17.2c

Note: *This procedure applies to the automatic release mechanism used on all models except the 2006 and later TRX250EX. The cable and release mechanism used on the 2006 and later TRX250EX are covered at the end of this Section.*

1   Remove the front crankcase cover (see Section 16).

2   Unscrew the clutch adjusting nut completely and remove its washer (see Chapter 1). Remove the release mechanism components from the crankcase and the cover **(see illustrations)**.

17.2a Slide the thrust washer and lifter lever (arrows) off the shaft (this is a 350) . . .

17.2b . . . and this is a 250

17.2c Remove the ball retainer, noting how the spring fits on it

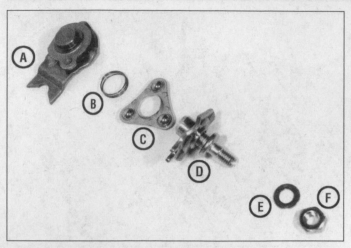

17.3a Release mechanism details

A	Lifter cam	D	Adjuster bolt
B	Spring	E	Washer
C	Ball retainer	F	Adjuster bolt locknut

Inspection

Refer to illustration 17.3a and 17.3b

3   Check for visible wear or damage at the contact points of the lever and cam and the friction points of the cam, ball retainer and lifter

17.3b Check the clutch lever for a worn roller (left arrow) and damaged splines (right arrow)

plate **(see illustrations)**. Check the spring for bending or distortion. Replace any parts that show problems. Replace the lifter plate O-ring in the crankcase cover whenever it's removed.

Installation

Refer to illustrations 17.4a and 17.4b

4   Installation is the reverse of the removal steps, with the following additions:

a)   *On 350 models, make sure the lifter plate engages the pin inside the crankcase cover (see illustration). On 250 models, install the O-ring on the adjusting plate screw and make sure the adjusting plate pin fits in the small hole in the cover.*

17.4a When installing a 350 lifter cam, don't forget the O-ring (right arrow) and position the lifter cam over the pin (left arrow)

17.4b Align the punch mark on the lever (left arrow) with the wide tooth of the spindle washer (right arrow)

17.7a Hold the clutch with a tool like this one . . .

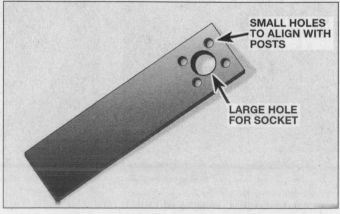

SMALL HOLES TO ALIGN WITH POSTS

LARGE HOLE FOR SOCKET

17.7b . . . or you can make a holding tool from flat steel stock

17.8a Remove the four bolts and the lifter plate (arrow) . . .

17.8b . . . and remove the pressure plate springs

17.8c Grind or file away the staked portion of the locknut (arrow), then unscrew it

b) *Align the wide groove on the shaft with the wide tooth of the spindle washer and make sure the lever engages the cam* **(see illustration)**.

5   Refill the engine oil and adjust the clutch (see Chapter 1).

Clutch
Removal
Refer to illustrations 17.7a, 17.7b and 17.8a through 17.8i

6   Remove the front crankcase cover and the centrifugal clutch (see Section 15). Remove the release mechanism as described above.

7   Note that if an air wrench is not available, the clutch needs to be prevented from rotating while removing the locknut. The Honda service tool (part no. 07HGB-001010A or 07HGB-001010B) resembles a steering wheel puller and is attached to the spring posts on the pressure plate **(see illustration)**. It is then held with a breaker bar while the nut is loosened with a socket. A similar tool can be fabricated from metal stock **(see illustration)**.

8   Refer to the accompanying illustrations to remove the clutch components **(see illustrations)**.

17.8d Pull off the clutch center . . .

17.8e . . . remove the outer friction plate . . .

17.8f . . . then the metal plate, then remove the remaining friction and metal plates . . .

17.8g Remove the pressure plate from the clutch housing . . .

17.8h . . . then remove the thrust washer and clutch housing . . .

17.8i . . . and the clutch outer guide

17.9 Clutch inspection points

A *Pressure plate posts*
B *Pressure plate friction surface*
C *Clutch housing slots*
D *Clutch housing bearing surface*
E *Primary driven gear*

Inspection

Refer to illustrations 17.9, 17.11, 17.12, 17.13, 17.16 and 17.17

9   Check the bolt posts and the friction surface on the pressure plate for damaged threads, scoring or wear **(see illustration)**. Replace the pressure plate if any defects are found.

10   Check the edges of the slots in the clutch housing for indentations made by the friction plate tabs **(see illustration 16.9)**. If the indentations are deep they can prevent clutch release, so the housing should be replaced with a new one. If the indentations can be removed easily with a file, the life of the housing can be prolonged to an extent. Also, check the driven gear teeth for cracks, chips and excessive wear and the springs on the back side for breakage. If the gear is worn or damaged or the springs are broken, the clutch housing must be replaced with a new one. Check the bearing surface in the center of the clutch housing for score marks, scratches and excessive wear.

11   Measure the free length of the clutch springs **(see illustration)** and compare the results to this Chapter's Specifications. If the springs have sagged, or if cracks are noted, replace them with new ones as a set.

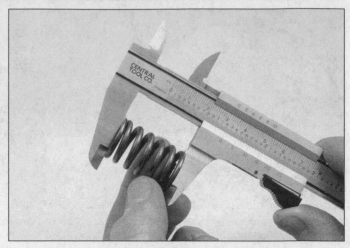

17.11 Measure the clutch spring free length

12   If the lining material of the friction plates smells burnt or if it is glazed, new parts are required. If the metal clutch plates are scored or discolored, they must be replaced with new ones. Measure the thickness of the friction plates **(see illustration)** and replace with new parts any friction plates that are worn.

13   Lay the metal plates, one at a time, on a perfectly flat surface (such as a piece of plate glass) and check for warpage by trying to slip a feeler gauge between the flat surface and the plate **(see illustration)**. The feeler gauge should be the same thickness as the maximum warp listed in this Chapter's Specifications. Do this at several places around the plate's circumference. If the feeler gauge can be slipped under the plate, it is warped and should be replaced with a new one.

14   Check the tabs on the friction plates for excessive wear and mushroomed edges. They can be cleaned up with a file if the deformation is not severe. Check the friction plates for warpage as described in Step 14.

15   Check the clutch outer guide for score marks, heat discoloration and evidence of excessive wear **(see illustration 17.8i)**. Measure its inner and outer diameter and compare them to the values listed in this Chapter's Specifications (a dealer can do this if you don't have precision measuring equipment). Replace the outer guide if it's worn beyond the specified limits. Also measure the end of the transmission mainshaft where the outer guide rides; if it's worn to less than the limit listed in this Chapter's Specifications, replace the mainshaft (Section 26).

16   Check the clutch lifter plate for wear and damage. Rotate the inner race of the bearing and check for roughness, looseness or excessive noise **(see illustration)**.

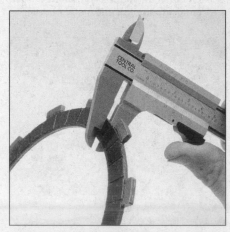

17.12 Measure the thickness of the friction plates

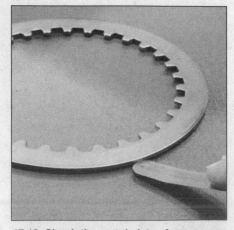

17.13 Check the metal plates for warpage

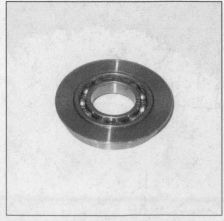

17.16 Check the ball bearing in the center of the lifter plate for roughness, looseness or noise

17.17 Check the clutch center splines for wear or damage

17.20a Install a friction plate first, then a metal plate, then alternate the remaining friction and metal plates . . .

17.20b . . .a friction plate goes on last

17   Check the splines of the clutch center for wear or damage and replace the clutch center if problems are found **(see illustration)**.

Installation

Refer to illustrations 17.20a, 17.20b, 17.21a, 17.21b, 17.24 and 17.25

18   Lubricate the inner and outer surfaces of the clutch outer guide with moly-based grease and install it on the crankshaft.

19   Install the clutch housing and thrust washer on the mainshaft **(see illustration 17.8h)**.

20   Coat the friction plates with engine oil. Install a friction plate on the disc, followed by a metal plate, then alternate the remaining friction and metal plates until they're all installed (there are six friction plates and five metal plates). Friction plates go on first and last, so the friction material contacts the metal surfaces of the clutch center and the pressure plate **(see illustrations)**.

21   Install the clutch center over the posts, then install the assembly in the clutch housing **(see illustrations)**.

22   Install the washer on the mainshaft **(see illustration 17.8d)**.

23   Coat the threads of a new locknut with clean engine oil and install it on the mainshaft. Hold the clutch with one of the methods described in Step 7 and tighten the locknut to the torque listed in this Chapter's Specifications.

24   Stake the locknut with a hammer and punch **(see illustration)**.

25   Install the clutch springs and the lifter plate **(see illustration)**.

17.21a Install the clutch center . . .

Tighten the bolts next to the grooves first, then tighten the other two bolts to the torque listed in this Chapter's Specifications.

26   The remainder of installation is the reverse of the removal steps.

17.21b . . . then install the assembly on the mainshaft

17.24 Stake the new locknut

17.25 Install the springs and lifter plate and install the bearing in the lifter plate if it was removed

17.35 Pull back the lever to relieve tension on the lifter piece and pull the lifter piece out of the crankcase cover

17.38 The lifter piece engages the lever notch like this when they're installed in the cover

17.39a Inspect the seal and bearing in the top of the cover . . .

Cable and release mechanism (2006 and later TRX250EX)

Cable

27   Loosen the locknut and adjustment nut, then unbolt the clutch cable bracket from the engine.

28   Disengage the clutch cable from the lifter arm clevis. **Note:** *If necessary, pry open the lifter arm gap slightly with a screwdriver so the cable can pass through it.*

29   If necessary, loosen the locknut and adjustment nut some more to create enough slack to disengage the cable from the engine bracket. Slip the cable out of the bracket.

30   Up at the handlebars, pull back the cable rubber dust boot, then back off the locknut and adjuster (see the clutch cable adjustment procedure in Chapter 1). Align the slots in the adjuster and locknut, then turn the cable so it aligns with the slot in the underside of the lever and lower the cable end plug out of the lever.

31   Note the routing of the cable and remove it. **Note:** *It's a good idea to attach a piece of string to one end of the cable before removing it; that way, when the cable is pulled out, the string will be in the cable's routing path so you can route the new cable correctly.*

32   Installation is the reverse of removal. Make sure the cable is routed correctly, with no kinks.

33   Adjust the clutch cable freeplay (see Chapter 1).

Release mechanism

Refer to illustration 17.35

34   Remove the front crankcase cover (see Section 16).

35   Remove the clutch lifter piece from the lifter arm **(see illustration)**. If it doesn't come out easily, twist the lifter arm clockwise to relieve the spring tension.

36   Note how the straight (lower) end of the spring fits against the cover, then carefully detach the hooked (upper) end of the spring from the pin in the lifter arm shaft. Tap the spring pin into the shaft with a hammer and punch until the pin is flush with the surface of the shaft. This will keep the pin from catching on the bearing and seal when you pull the lifter arm out of the cover.

37   Pull the lifter arm out of the front crankcase cover and set aside the parts for the lifter arm assembly for later inspection. Store the lifter piece, arm and return spring in a plastic bag so you don't misplace them.

Inspection

Refer to illustrations 17.38, 17.39a and 17.39b

38   Inspect the lifter arm and the clutch lifter piece **(see illustration)** for damage and wear. Make sure the lifter arm isn't bent. The clutch

17.39b . . . and the bearing inside the cover

lifter piece should slide in and out of its bore in the front crankcase cover without binding. Inspect the return spring for fatigue and other damage. If any part of the lifter arm assembly is damaged, replace it.

39   Inspect the lifter arm seal and bearings in the right crankcase cover **(see illustrations)**. If the seal is cracked, torn or worn, remove it and install a new seal. Replace the bearings if their condition is in doubt or if you can see and wear or damage.

40   Inspect the lifter plate bearing on the change clutch for wear and damage **(see illustration 17.8a)**. Turn the bearing inner race with your finger; it should turn smoothly and quietly. Make sure that the bearing outer race fits tightly into the lifter plate. If the bearing is damaged or worn, replace the lifter plate and bearing as an assembly.

Installation

41   Install the lifter bearing on the clutch (if it was removed) **(see illustration 17.8a)**.

42   Place the spring in its installed position in the front cover so its hooked end will be upward when the cover is installed. Slide the lifter arm shaft into its bore, taking care not to damage the seal.

43   Using a 3 mm (1/8-inch) punch, tap the spring pin from the chamfered side of its bore until the pin protrudes 3 mm (1/8 inch) from the shaft.

44   Place the curved end of the spring over its cast rib in the cover, then slip the hooked end of the spring over the pin.

45   Pull back on the lifter arm until the lifter piece notch aligns with the bore, then install the lifter piece.

18.2a Slide the primary drive gear off . . .

18.2b . . . and remove the thrust washer (arrow)

46   Install the lifter arm and its spring, then install the clutch lifter piece, engaging it with the lifter arm **(see illustrations 17.35 and 17.38)**.
47   Install the right crankcase cover and install its bolts finger tight. Don't forget to reattach the clutch cable bracket.
48   Tighten the bolts in stages, using a criss-cross pattern, to the torque listed in this Chapter's Specifications.
49   Reattach the clutch cable to the clutch lifter arm by reversing the procedure in Step 28.

18   Primary drive gear (350 models) - removal, inspection and installation

Removal
Refer to illustrations 18.2a and 18.2b
1   Remove the centrifugal clutch and change clutch (Sections 15 and 16).
2   Slide the primary drive gear off the crankshaft and remove the washer **(see illustrations)**.

Inspection
Refer to illustrations 18.4 and 18.5
3   Check the drive gear for obvious damage such as a worn inner

bushing, damaged splines and chipped or broken teeth. Replace it if any of these problems are found.
4   Measure the inside diameter of the bushing at the outer end and inner end of the gear **(see illustration)**. A Honda dealer or machine shop can do this if you don't have precision measuring equipment. If either bushing is worn beyond the diameter listed in this Chapter's Specifications, replace the gear.
5   Measure the crankshaft at the two points where the bushings ride **(see illustration)**. If either point is worn to less than the diameter listed in this Chapter's Specifications, replace the crankshaft (Section 28).

Installation
6   Installation is the reverse of the removal steps.

19   Oil pump and cooler - removal, inspection and installation

Oil pump
Removal (350 models)
Refer to illustrations 19.2a, 19.2b, 19.3 and 19.4
1   Remove the front crankcase cover (see Section 16).

18.4 Measure the diameter of the bushings (arrow); there's one inside each end of the gear

18.5 Measure the points on the crankshaft where the bushings ride (arrows)

19.2a On 350 models, unbolt the feed pipe retainer (arrow) . . .

19.2b . . . and work the feed pipe and O-rings free of the crankcase

19.3 Unbolt the scavenge pipe, then work the pipe and O-rings free of the crankcase

19.4 Unbolt the oil pump (arrows) and take it off the engine (350)

2   Unbolt the feed pipe retainer **(see illustration)**. Work the feed pipe and O-rings free of the crankcase **(see illustration)**.
3   Unbolt the scavenge pipe. Work the pipe and O-rings free of the crankcase **(see illustration)**.

4   Remove the pump mounting bolts and take the pump off the engine **(see illustration)**.

Removal (250 models)
Refer to illustrations 19.7, 19.8a and 19.8b
5   Remove the front crankcase cover and centrifugal clutch (see Section 16).
6   Remove the cam chain tensioner and slider (see Section 20).
7   Remove the snap-ring from the oil pump sprocket, then slide the sprocket off the oil pump **(see illustration)**.
8   Remove the pump mounting bolts and take the pump off the engine **(see illustrations)**.

Inspection
Refer to illustrations 19.9a, 19.9b, 19.10a, 19.10b, 19.14a, 19.14b and 19.14c
9   If you're working on a 350 model, remove the bolt **(see illustration)**. Separate the pump base and body from the spacer. Remove the rotors and shaft **(see illustration)**.
10   If you're working on a 250 model, remove the shaft snap-ring and washer, as well as the screw that secures the side plate to the pump body **(see illustration)**. Remove the side plate, rotors and shaft **(see illustration)**.
11   Wash all the components in solvent, then dry them off. Check the

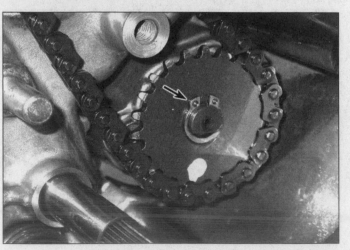

19.7 Remove the snap-ring (arrow) and slide the sprocket off the oil pump (250)

19.8a Unbolt the oil pump and take it off the engine (250) . . .

19.8b and locate the pump dowels (arrows)

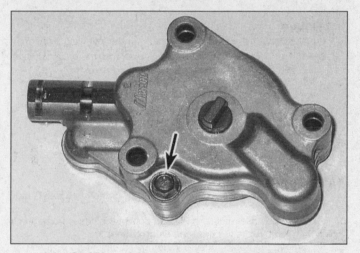

19.9a On 350 models, remove the cover bolt (arrow) . . .

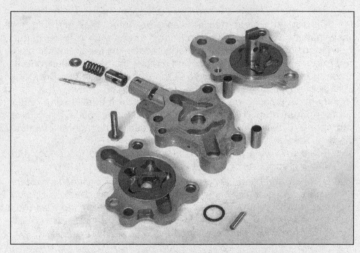

19.9b . . . and disassemble the pump

pump body, the rotors and the cover for scoring and wear. If any damage or uneven or excessive wear is evident, replace the pump. If you are rebuilding the engine, it's a good idea to install a new oil pump.

12　If you're working on a 350 model, place the feed rotors (thin)

rotors in the pump body. Place the scavenge rotors (thick) in the pump base.

13　If you're working on a 250 model, place the pump rotors in the pump body.

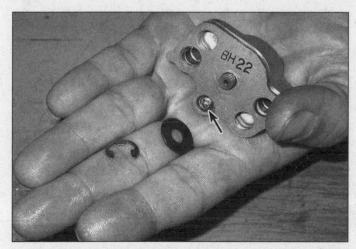

19.10a On 250 models, remove the shaft snap-ring, thrust washer and screw (arrow) . . .

19.10b . . . then remove the side plate to expose the rotors

19.14a Use a feeler gauge to measure the inner-to-outer rotor clearance

19.14b . . . and the outer rotor-to-body clearance

19.14c Lay a straightedge across the rotors and pump body and measure the gap with a feeler gauge

14   Measure the clearance between the outer rotor and body or base, and between the inner and outer rotors, with a feeler gauge **(see illustrations)**. Place a straightedge across the pump body or base and rotors and measure the gap with a feeler gauge **(see illustration)**. If any of the clearances are beyond the limits listed in this Chapter's Specifications, replace the pump.

15   If you're working on a 350 model, remove the relief valve cotter pin **(see illustration 19.9b)**. Dump out the stopper plate, relief valve and spring. Check the components for wear or damage and replace them if necessary.

16   Reassemble the pump by reversing the disassembly steps, with the following additions:

 a) *Before assembling the pump, pack the cavities between the rotors with petroleum jelly - this will ensure the pump develops suction quickly and begins oil circulation as soon as the engine is started.*

 b) *Make sure the dowels are in place. On 350 models, also make sure both pump drive pins and the thrust washer are in place.*

 c) *Make sure there's no gap between the pump body, spacer and base 350) or pump body and side plate (250). Tighten the pump bolt or screw securely, but don't overtighten it and strip the threads.*

 d) *On 350 models, install the relief valve spring stopper with its small end toward the spring. Use a new cotter pin to secure the relief valve.*

 e) *On 250 models, install the washer and snap-ring on the pump shaft.*

Installation

17   Installation is the reverse of removal, with the following additions:

 a) *On 350 models, engage the oil pump shaft with the balancer shaft when you install the pump on the engine.*

 b) *Tighten the oil pump bolts securely, but don't overtighten them and strip the threads.*

 c) *Use new O-rings on the feed and scavenge pipes. Coat the O-rings with clean engine oil.*

Oil cooler

Removal

Refer to illustrations 19.19a, 19.19b, 19.20a and 19.20b

18   Drain the engine oil and remove the inner front fenders (see Chapters 1 and 7).

19   Remove the hose retainer bolts and detach the hoses from the oil cooler, then remove the O-rings **(see illustrations)**.

20   To remove the hoses, unscrew the hose retaining bolts, then pull the hoses out of the front crankcase cover and remove the O-rings **(see illustrations)**.

Inspection

21   Check the oil cooler for signs of leakage, especially at the hose joints. It's a good idea to replace the O-rings whenever they're removed, but they should be definitely be replaced if they're compressed, brittle or deteriorated.

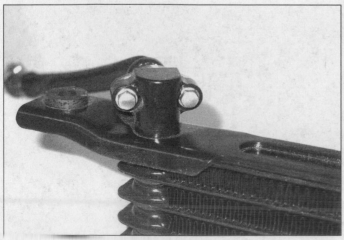

19.19a Remove the hose retainer bolts and detach the hoses from the oil cooler . . .

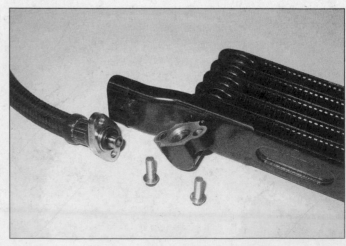

19.19b . . . then remove the O-rings

19.20a Unscrew the hose retaining bolts, then pull the hoses out of the front crankcase cover and remove the O-rings (this is a 350) . . .

19.20b . . . and here are the hoses on a 250

20.2a Tensioner and slider mounting (350 models)

20.2b Tensioner and slider mounting (250 models)

22   Check the hoses for deterioration, and look closely for leakage or cracks at the joints where the hose meets the metal fitting. Replace the hoses if there are any doubts about their condition - a hose failure will cause a sudden catastrophic loss of oil pressure.
23   Check the oil cooler fins for clogging or bending. Clean out any debris with a garden hose or low-pressure compressed air. Carefully straighten any bent fins with a small screwdriver, taking care not to puncture the oil passages.

Installation

24   Installation is the reverse of the removal steps. Use new O-rings and tighten the mounting bolts securely, but don't overtighten them and strip the threads.

20 Cam chain tensioner - removal and installation

Removal

Refer to illustrations 20.2a, 20.2b and 20.2c

1   Remove the cylinder (see Section 11). If you're working on a 350 model, remove the alternator and starter drive gear (see Chapter 8). If you're working on a 250 model, remove both clutches (see Sections 16 and 17).
2   Unbolt the tensioner from the crankcase **(see illustrations)**. After you remove it, squeeze the stopper, push in the tensioner piston and

secure it with a straightened paper clip or similar tool **(see illustration)**.
3   If you need to remove the tensioner slider, unbolt it **(see illustration 20.2a or 20.2b)**, then take the slider off and remove the washer.

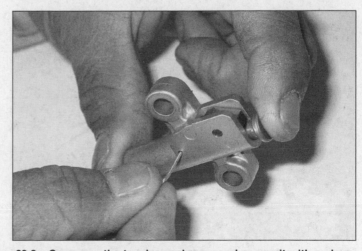

20.2c Compress the tensioner plunger and secure it with a piece of wire

20.4 Tensioner and slider details (350 shown; 250 similar)

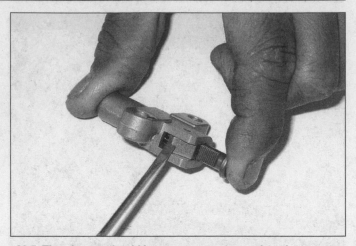

20.5 The plunger should be easy to compress when the stopper block pressed in

21.4a On 350 models, the keyway should be straight up and the camshaft sprocket punch mark should align with the cast pointer (arrows); on 2002 and later models, use the punch mark next to the 350 mark

21.4b On 1997 through 2001 250 models, align the punch mark with the case pointer . . .

Inspection

Refer to illustrations 20.4 and 20.5

4   Check the friction surfaces on the tensioner and slider for wear or damage **(see illustration)**. Replace them if problems are found.

21.4c . . . on 2002 and later 250 models, use the punch mark next to the 250 mark (arrows)

5   Try to push the piston into the tensioner by hand. It should not go. When the stopper block is pressed into the tensioner, it should be easy to push the piston in **(see illustration)**. If the tensioner doesn't perform as described, replace it.

Installation

6   Installation is the reverse of the removal steps. Tighten the slider bolt to the torque listed in this Chapter's Specifications.

---

21   Camshaft, chain and sprockets - removal, inspection and installation

---

Removal

Refer to illustrations 21.4a, 21.4b, 21.4c, 21.5a and 21.5b

1   Refer to the valve adjustment procedure in Chapter 1 and position the piston at top dead center (TDC) on its compression stroke.

2   Remove the valve cover, rocker arms, pushrods, cylinder head, cylinder and valve lifters (see Sections 7, 8, 11 and 14).

3   Remove the alternator rotor (see Chapter 8).

4   Verify that the crankshaft and piston are still at TDC on the compression stroke. When this occurs, the crankshaft keyway will be straight up (in a line with the cylinder studs). In addition, the dimple on

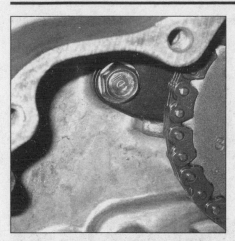

21.5a Unbolt the camshaft retainer (250 shown) . . .

21.5b . . . and pull the camshaft out (350 shown)

21.7a Spin the camshaft bearing and check it for roughness, looseness and noise

the camshaft sprocket will align with the cast mark on the crankcase **(see illustrations)**. **Note:** *On 2002 and later Rancher and Recon models and all TRX250EX models, there are two dimples, one next to a 350 mark and one next to a 250 mark. Use the dimple next to the 350 mark if you're working on a 350 model, or use the dimple next to the 250 mark if you're working on a 250 model.*

5   Unbolt the camshaft bearing retainer **(see illustration)**. Pull the camshaft out of the engine, disengaging the chain from the crankshaft as you do so **(see illustration)**.

Inspection

Refer to illustrations 21.7a, 21.7b, 21.7c and 21.9

Note: *Before replacing the camshaft or crankcase because of damage, check with local machine shops specializing in ATV or motorcycle engine work. In the case of the camshaft, it may be possible for cam lobes to be welded, reground and hardened, at a cost far lower than that of a new camshaft. If the bearing surfaces in the crankcase are damaged, it may be possible for them to be bored out to accept bearing inserts. Due to the cost of a new crankcase it is recommended that all options be explored before condemning it as trash!*

6   On 350 models, inspect the cam bearing surfaces inside the crankcase, using a flashlight if necessary. Look for score marks, deep scratches and evidence of galling (a pitted appearance).

7   Check the camshaft lobes for heat discoloration (blue appear-

ance), score marks, chipped areas, flat spots and galling **(see illustrations)**. Measure the height of each lobe with a micrometer **(see illustration)** and compare the results to the minimum lobe height listed in this Chapter's Specifications. If damage is noted or wear is excessive, the camshaft must be replaced. Also, be sure to check the condition of the valve lifters and their bores as described in Section 14.

8   Spin the ball bearing on each end of the camshaft and check for roughness, looseness and noise **(see illustration 21.7a)**. Note that on 250 models, the bearing on the end of the camshaft opposite the sprocket will remain in the crankcase. If there's any problem with a bearing mounted on the camshaft, replace the camshaft and bearing(s) as an assembly. On 250 models, if there's a problem with the camshaft bearing that remained in the crankcase, disassemble the crankcase as described in Section 24 to replace the bearing.

9   If you're working on a 350 model, check the operation of the compression release mechanism. Hold the camshaft with the base of the exhaust lobe upright **(see illustration)**. Press on the exhaust lobe. Pressing on the left side should cause the decompressor lobe to lock above the level of the exhaust lobe base. Pressing on the right side should lower the decompressor lobe below the level of the exhaust lobe base.

10   Except in cases of oil starvation, the camshaft chain wears very little. If the chain has stretched excessively, which makes it difficult to maintain proper tension, replace it with a new one.

21.7b Check the cam lobes for wear - here's a good example of damage which will require replacement (or repair) of the camshaft

21.7c Measure the height of the cam lobes with a micrometer

21.9 If you're working on a 350 model, check the operation of the compression release mechanism (see text)

22.1a On 350 models, remove the bolt (arrow) and detach the bracket . . .

22.1b . . . then rotate the cable end to align with the slot in the lever (arrow) and slip it out

11   Check the sprocket for wear, cracks and other damage. If problems are found, replace the sprocket and camshaft as an assembly. If the sprocket is worn, the chain is also worn, and possibly the sprocket on the crankshaft. If wear this severe is apparent, the entire engine should be disassembled for inspection.
12   Check the chain tensioner and slider (see Section 20). If they are worn or damaged, replace them.

Installation

13   If you're working on a 350 model, make sure the crankshaft keyway is still straight up **(see illustration 21.4a)**. If you're working on a 250 model, make sure the T mark on the alternator rotor is at the Top Dead Center position (see the valve adjustment procedure in Chapter 1).
14   Lubricate the cam bearings with engine oil and the lobes with molybdenum disulfide grease. Place the chain on the cam sprocket and place the camshaft in its bores. Turn the camshaft so its alignment mark is positioned correctly, then engage the chain with the crankshaft sprocket **(see illustration 21.4a, 21.4b or 21.4c)**. Recheck to make sure the timing marks are lined up correctly after the chain is installed on both sprockets.
15   Install the bearing retainer **(see illustration 21.5a or 21.5b)**. Apply non-permanent thread locking agent to the threads of the retainer bolt and tighten it to the torque listed in this Chapter's Specifications.
16   Recheck the position of the crankshaft keyway and the timing

22.2 On 250 models, unbolt the cover and bracket (arrows), then rotate the cable end to align with the slot in the lever and slip it out

marks on the sprocket and crankcase **(see illustration 21.4a, 21.4b or 21.4c)**. **Caution:** *Don't run the engine with the marks out of alignment or severe engine damage could occur.*
17   The remainder of installation is the reverse of removal. Refer to Chapter 1 and adjust the valve clearances.

22 Reverse lock mechanism - cable replacement, removal, inspection and installation

Note: *This section covers the external components of the reverse lock mechanism. To service the reverse stopper arm, located inside the alternator cover, refer to Section 24.*

Cable replacement

Refer to illustrations 22.1a, 22.1b and 22.2
1   If you're working on a 350 model, unbolt the cable bracket from the engine, then rotate the cable to align it with the slot in the lever and slip the cable out of the lever **(see illustrations)**.
2   If you're working on a 250 model, unbolt the cable end cover and remove the cable bracket bolt **(see illustration)**. Rotate the cable to align it with the slot in the lever and slip the cable out of the lever.
3   Pull the cable housing out of the handlebar bracket. Rotate the cable out of the bracket, then align the cable with the slot in the bracket and slip the cable end down out of the bracket.
4   Note the location of the cable along the frame. Detach it from its retainers and install the new cable in the same location, making sure to secure it with all of the retainers.
5   Reconnect the ends of the cable at the handlebar, lower bracket and lever. Refer to Chapter 1 and adjust the cable.

Removal

6   Disconnect the reverse cable as described above.
7   Remove the nut or bolt and take the washer and lever off the shaft **(see illustration 22.1b or 22.2)**.

Inspection

8   Check the shaft oil seal for leaks. If necessary, pry it out, taking care not to damage the seal bore. Coat the lip of a new seal with clean engine oil and tap it into place, lip toward the engine, using a socket the same diameter as the seal.

Installation

9   Installation is the reverse of the removal steps, with the following addition: Install the reverse stopper arm with its slit mark facing away from the engine. Tighten its nut securely, but don't overtighten and strip the threads.

23.2 Unbolt the wiring harness retainer (350 models, if equipped) and recoil starter (350 shown)

23.3 Remove the bolt and lift off the ratchet cover

23 Recoil starter - removal, inspection and installation

Removal

Refer to illustration 23.2

1 Remove the air cleaner from the vehicle (see Chapter 3).

2 Remove the bolts that secure the wiring harness retainer (350 models), if equipped, and recoil starter to the rear of the engine **(see illustration)**. Take the recoil starter off.

Inspection

Refer to illustrations 23.3, 23.4, 23.5, 23.6, and 23.7

3 Remove the bolt and lift off the ratchet cover **(see illustration)**.

4 Note how the ratchet guide, spring seat, ratchet and pin are installed, then lift them off the starter **(see illustration)**.

5 Untie the knot in the starter rope. Carefully release the rope into the starter and remove the pull handle **(see illustration)**.

6 Note how the rope is wound onto the pulley (on 350 models, counterclockwise, viewed from the ratchet side; on 250 models, clockwise, viewed from the ratchet side), then unwind it and pull the knot out of the pocket **(see illustration)**.

7 **Warning:** *Whenever you handle the recoil spring, wear eye protection and heavy gloves to avoid injury if the spring uncoils suddenly. If*

23.4 Lift off the ratchet guide, spring seat, ratchet and pin

the spring is in good condition, leave it installed. If it's broken, carefully remove it from the cover **(see illustration)**.

8 Check all other parts for wear or damage and replace worn or damaged parts.

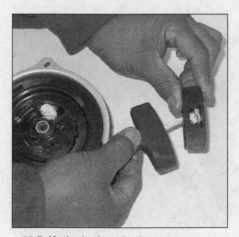

23.5 Untie the knot in the starter rope, release the rope into the starter and remove the pull handle

23.6 The rope is wound onto the pulley counterclockwise on 350 models and clockwise on 250 models, viewed from the ratchet side; unwind it and pull the knot out of the pocket

23.7 If the spring is broken, carefully remove it from the cover

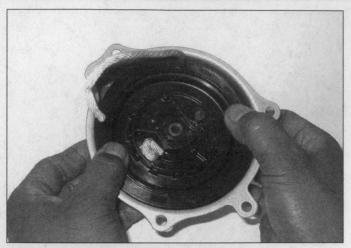

23.11 Guide the end of the rope through the cover hole and pull handle, then tie a square knot in the end of the rope to secure it

24.8 Slide the gearshift plate off the sub-gearshift spindle, noting how the pin fits between the ends of the spring

Installation

Refer to illustration 23.11

9   If you removed the spring, reinstall it, hooking the end into the notch in the cover **(see illustration 23.7)**.

10   Lubricate the pulley shaft in the cover with multi-purpose grease. Install the pulley in the cover, making sure the pulley engages the inner end of the recoil spring.

11   Place the rope in the pulley notch, then turn the pulley 4 turns counterclockwise (350 models) or clockwise (250 models) to preload the recoil spring **(see illustration)**. Guide the end of the rope through the cover hole and pull handle, then tie a square knot in the end of the rope to secure it.

12   Lubricate the pin and ratchet with multi-purpose grease, then install them **(see illustration 23.4)**. Install the spring, then install the spring seat and ratchet guide over them.

13   Install the ratchet cover **(see illustration 23.3)**. Lubricate the threads of the ratchet cover bolt with engine oil, then install it. Tighten the bolt securely, but don't overtighten it and strip the threads.

14   Hold the cover down on a work surface and pull the rope two or three times to make sure the recoil mechanism works properly.

15   The remainder of installation is the reverse of the removal steps. Tighten the recoil starter bolts securely, but don't overtighten them and strip the threads.

---

24 External shift mechanism - removal, inspection and installation

1   The external shift mechanism includes the shift pedal (foot shift models), gearshift plate, stopper arm and cam, gearshift spindle and reverse stopper arm. Service to the sub-gearshift spindle requires disassembling the crankcase and is covered in Sections 25 through 27.

Shift pedal

2   The shift pedal is used on all except electric shift models.

Removal

3   Look for alignment marks on the end of the shift pedal and shift shaft. If they aren't visible, make your own marks with a sharp punch.

4   Remove the shift pedal pinch bolt and slide the pedal off the shaft.

Inspection

5   Check the shift pedal for wear or damage such as bending. Check the splines on the shift pedal and shaft for stripping or step wear. Replace the pedal or shaft if these problems are found.

Installation

6   Install the shift pedal. Line up its punch marks and tighten the pinch bolt to the torque listed in this Chapter's Specifications.

Gearshift plate, stopper arm and cam (350 models)

Removal

Refer to illustrations 24.8, 24.9 and 24.10

7   Remove the change clutch (see Section 16) and the oil pipes (see Section 19).

8   Slide the gearshift plate off the sub-gearshift spindle, noting how the pin fits between the ends of the spring **(see illustration)**.

9   Push the stopper arm away from the gearshift cam, then pull the gearshift cam out **(see illustration)**. Be careful not to lose the dowel.
Note: *The transmission is in Neutral when the dowel is aligned with the cast pointer on the crankcase.*

10   Note how the stopper arm spring fits against the crankcase and over the stopper arm **(see illustration)**. Unbolt the stopper arm and remove it together with the spring.

Inspection

Refer to illustration 24.14

11   Check the gearshift cam and replace it if it's worn or damaged.

12   Check the sub-gearshift spindle for bends and damage to the

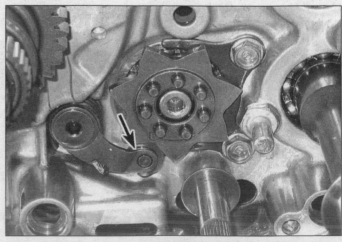

24.9 Push the stopper arm (arrow) away from the gearshift cam and pull the cam out; don't lose the dowel

24.10 Note how the stopper arm spring hooks over the stopper arm and rests against the crankcase, then remove the stopper arm and spring

24.14 Inspect the gearshift plate, return spring and pawl spring

splines. If problems are found, refer to Sections 25 through 27 for removal procedures.

13   Make sure the return spring pin isn't loose **(see illustration 24.8)**. If it is, unscrew it, apply a non-hardening locking compound to the threads, then reinstall it and tighten it to the torque listed in this Chapter's Specifications.

14   Check the gearshift plate for wear or damage **(see illustration)**. Check the return spring (heavy) and pawl spring (light) for bending or weakness. If problems are found, replace the gearshift plate as an assembly.

Installation

Refer to illustrations 24.17a and 24.17b

15   Place the return spring on the stopper arm **(see illustration 24.10)**. Apply non-permanent thread locking agent to the threads of the stopper arm bolt, then install the stopper arm and tighten the bolt to the torque listed in this Chapter's Specifications. Make sure the end of the return spring is positioned against the crankcase.

16   Install the dowel pin in its hole in the shift drum, then pull the stopper arm out of the way and place the gearshift cam over the dowel. Apply non-permanent thread locking agent to the threads of the gearshift cam bolt, then install it and tighten it to the torque listed in this Chapter's Specifications.

17   Install the gearshift plate, aligning its wide spline with the matching spline on the sub-gearshift spindle **(see illustration)**. Position the ends of the return spring on each side of the return spring pin **(see illustration)**.

18   The remainder of installation is the reverse of the removal steps.

19   Refill the engine oil (see Chapter 1).

Gearshift spindle (350 models)

Removal

Refer to illustration 24.21

20   Remove the alternator cover and rear crankcase cover (see Chapter 8).

21   Unscrew the spindle retaining bolt **(see illustration)**. Pull the spindle out of the arm, then remove the arm from the engine.

Inspection

Refer to illustration 24.24

22   Check the splines of the arm and spindle for wear or damage. Replace them both if problems are found.

23   Check the gearshift arm for wear at the tip and replace it if problems are found.

24.17a Align the gearshift plate's wide spline with the matching spline on the sub-gearshift spindle

24.17b Position the ends of the return spring on each side of the return spring pin

24.21 Remove the spindle bolt (right arrow), pull it out of the arm (left arrow) and remove the arm

24.24 Check the spindle oil seal for leaks

24.26 Align the wide spline in the spindle with the punch mark on the arm

24   Check the spindle oil seal for signs of leakage **(see illustration)**. If necessary, pry it out of the crankcase and tap in a new one with a seal driver or a socket the same diameter as the seal. The lip of the seal faces into the engine.

Installation
Refer to illustration 24.26

25   Lubricate the lip of the spindle seal with clean engine oil.

26   Position the gearshift arm in the crankcase, then start the spindle into the seal, taking care not to damage the seal. Align the wide spline in the spindle with the punch mark on the arm **(see illustration)**, then push the spindle into the arm.

27   Align the bolt holes in arm and spindle. Apply non-permanent thread locking agent to the bolt threads, then install the bolt and tighten it securely, but don't overtighten it and strip the threads.

Reverse stopper arm (350 models)
Removal
Refer to illustration 24.29

28   Remove the rear crankcase cover (see Chapter 8).

29   Disengage the reverse stopper arm from the shift drum and pull it out of the crankcase **(see illustration)**.

Inspection
30   Check the spring for weakness or bending. Check the arm tips and threads for wear and damage. Replace the arm as an assembly if problems are found.

Installation
31   Make sure the end of the spring is engaged with its notch in the arm **(see illustration 24.29)**.

32   Lubricate the friction surface on the shaft with clean engine oil.

33   Slide the shaft into its bore, lifting the spring to clear the edge of the crankcase. Push the shaft all the way in and engage it with the shift drum.

34   The remainder of installation is the reverse of the removal steps.

Gearshift spindle (250 models)
Removal
Refer to illustration 24.38

35   Remove the gearshift pedal as described above.

36   Remove the rear crankcase cover, together with the gearshift spindle (see the alternator removal procedure in Chapter 8).

37   If you're working on a 2001 or earlier Recon, or a 2005 or earlier TRX250EX, bend back the tabs on the lockwasher that secures the gearshift A arm to the gearshift spindle. On later models, the gearshift A arm isn't bolted to the spindle.

38   Unscrew the spindle bolt, then pull the spindle out of the alternator cover **(see illustration)**.

39   Remove the gearshift A arm from the alternator cover. On 2004 and earlier foot shift Recons, and all TRX250EX models, there's a washer on each end of the gearshift A arm.

Inspection
40   Inspection is the same as for 350 models, described in Steps 24 through 26 above.

24.29 Disengage the reverse stopper arm (arrow) from the shift drum and pull it out of the crankcase

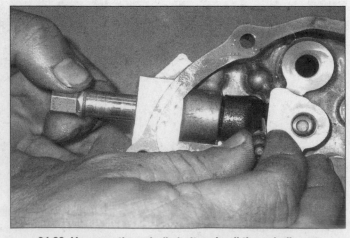

24.38 Unscrew the spindle bolt and pull the spindle out

24.42 Align the wide splines on spindle and arm

24.45 Slip the washer (if equipped) off the sub-gearshift spindle, then remove the spindle arm and slip off the washer behind it

Installation

Refer to illustration 24.42

41   Lubricate the lip of the spindle seal with clean engine oil.

42   Position the gearshift A arm (and its washers, if equipped) in the alternator cover **(see illustration 24.38)**. On models with washers, the washer with fingers goes in first, with the fingers positioned toward the gearshift A arm. Start the spindle into the seal, taking care not to damage the seal. Align the wide spline in the spindle with the wide spline in the gearshift A arm **(see illustration)**, then push the spindle into the gearshift A arm.

43   If the gearshift A arm is bolted to the spindle, align the bolt holes in arm and spindle. Install a new lockwasher, install the bolt and tighten it to the torque listed in this Chapter's Specifications. Bend the lockwasher tab up to secure the bolt.

Gearshift plate, stopper arm, cam and spindle (250 models)

Removal

Refer to illustrations 24.45, 24.46, 24.47, 24.48, 24.49, 24.50, 24.51a and 24.51b

44   Remove the alternator cover (see Chapter 8) and the change clutch (see Section 16).

45   Slip the washer (if equipped) off the sub-gearshift spindle, then remove the spindle arm and slip off the washer behind it **(see illustration)**.

46   Detach the left end of the master arm spring from the gearshift

24.46 Unhook the left end of the master arm spring from the gearshift guide plate

guide plate, noting how the right end of the spring is attached to the master arm **(see illustration)**.

47   Note how the return spring fits over its pin, then remove the return spring and master arm **(see illustration)**.

48   Unscrew the bolt from the gearshift cam plate **(see illustration)**.

24.47 Remove the return spring and master arm

24.48 Unbolt the gearshift cam plate, take the cam plate and guide plate off the cam and slip the thrust washer off the spindle

24.49 The 2005 Recon stopper arm (shown) pivots on the right side; on all other 250 models it pivots on the left side

24.50 Unbolt the stopper arm and remove it together with the spring

24.51a On the other side of the engine, pull the sub-gearshift spindle out of the engine . . .

Take the cam plate and guide plate off the cam, then slip the thrust washer off the spindle.

49   Pry the stopper arm away from the drum shifter and slip the drum shifter off the cam **(see illustration)**. **Note:** *The stopper arm on 2005 Recon models pivots on the right side of the shift drum. The stopper arm on all other 250 models pivots on the left side.* Locate the cam dowel pin; it may have come off with the cam or stayed in the shift drum.

50   Note how the stopper arm spring fits against the crankcase and over the stopper arm **(see illustration)**. Unbolt the stopper arm and remove it together with the spring.

51   On the other side of the engine, pull the sub-gearshift spindle out of the engine, then slip the thrust washer off the spindle **(see illustrations)**.

Inspection

52   Check the gearshift cam and replace it if it's worn or damaged.

53   Check the sub-gearshift spindle for bends and damage to the splines **(see illustration 24.51b)**. If problems are found, replace it.

54   Make sure the return spring pin isn't loose **(see illustration 24.45)**. If it is, unscrew it, apply a non-hardening locking compound to the threads, then reinstall it and tighten it to the torque listed in this Chapter's Specifications.

55   Check the gearshift linkage components for wear or damage.

Check the return spring (heavy) and master arm spring (light) for bending or weakness. If problems are found, replace the gearshift plate as an assembly.

Installation

Refer to illustrations 24.59, 24.61, 24.62 and 24.63

56   Install the thrust washer on the sub-gearshift spindle, then slide it into the engine **(see illustrations 24.51a and 24.51b)**.

57   Place the return spring on the stopper arm **(see illustration 24.50)**. Apply non-permanent thread locking agent to the threads of the stopper arm bolt, then install the stopper arm and tighten the bolt to the torque listed in this Chapter's Specifications. Make sure the end of the return spring is positioned against the crankcase.

58   Install the dowel pin in its hole in the shift drum, then pull the stopper arm out of the way and place the gearshift cam over the dowel. Apply non-permanent thread locking agent to the threads of the gearshift cam bolt, then install it and tighten it to the torque listed in this Chapter's Specifications.

59   Turn the drum shifter so its dowel pin aligns with the index mark (the raised rib on the crankcase) **(see illustration)**. This places the transmission in Neutral.

60   Slip the thrust washer onto the sub-gearshift spindle **(see illustration 24.48)**.

61   Place the gearshift plate on the engine, with its fingers straddling

24.51b . . . then slip the thrust washer off the spindle

24.59 Align the drum shifter dowel pin (left arrow) with the index mark (right arrow) to place the transmission in Neutral

24.61 When the transmission is in Neutral, the drum shifter keyway (right arrow) will be centered between the gearshift plate fingers (left arrow)

24.62 Place the gearshift cam plate key (right arrow) in the drum shifter keyway (left arrow)

the sub-gearshift spindle and its claws surrounding the two pins next to the drum shifter keyway **(see illustration)**.

62   Install the gearshift cam plate, making sure its key fits into the keyway in the center of the drum shifter **(see illustration)**. Apply non-permanent thread locking agent to the threads of the cam plate bolt, then tighten it to the torque listed in this Chapter's Specifications.

63   Install the master arm, positioning the ends of its spring around the pin **(see illustration)**. Hook the spring between the master arm and guide plate. Install the thrust washer on the sub-gearshift spindle.

64   Slide the arm onto the sub-gearshift spindle, aligning the wide splines on spindle and arm. Install the outer washer (if equipped).

65   The remainder of installation is the reverse of the removal steps.

Reverse stopper shaft (250 models)

Removal

Refer to illustration 24.67

66   Remove the rear crankcase cover (see Chapter 8).

67   Disengage the reverse stopper arm from the shift drum and pull the shaft out of the crankcase **(see illustration)**.

Inspection

68   Check the spring for weakness or bending. Check the arm tips and threads for wear and damage. Replace the arm as an assembly if problems are found.

Installation

69   Make sure the end of the spring is engaged with its notch in the arm **(see illustration 24.67)**.

70   Lubricate the friction surface on the shaft with clean engine oil.

71   Slide the shaft into its bore, lifting the spring to clear the edge of the crankcase. Push the shaft all the way in and engage it with the shift drum.

72   The remainder of installation is the reverse of the removal steps.

25   Crankcase - disassembly and reassembly

1   To examine and repair or replace the crankshaft, balancer (350 models), connecting rod, bearings and transmission components, the crankcase must be split into two parts.

Disassembly

Refer to illustrations 25.11, 25.13a, 25.13b, 25.14a, 25.14b and 25.16

2   Remove the engine from the vehicle (see Section 5).

3   Remove the carburetor (see Chapter 3).

4   Remove the alternator rotor, starter reduction gears and starter motor (see Chapter 8).

5   Remove the centrifugal clutch (see Section 17) and the change clutch (see Section 18).

24.63 Position the ends of the master arm spring (arrows) around the pin

24.67 Disengage the reverse stopper arm from the shift drum and pull the shaft out of the crankcase

25.11 Remove the output shaft gears (350 models)

25.13a Here are the 350 front crankcase bolts (arrows)

25.13b Here are the 250 front crankcase bolts (arrows)

9   Remove the camshaft (see Section 21).
10   Check carefully to make sure there aren't any remaining components that attach the front and rear halves of the crankcase together.
11   From the front side of the crankcase, remove the output shaft drive and driven gears (if equipped) **(see illustration)**.
12   Remove the dust seals from the engine hanger bushings if you haven't already done so. Drive the bushings out of the crankcase.
13   Loosen the rear crankcase bolts in two or three stages, in a criss-cross pattern **(see illustrations)**. Remove the bolts and label them; they are different lengths. If you're working on a 350 model, note the location of the side cover bracket.
14   Loosen the front crankcase bolts in two or three stages, in a criss-cross pattern **(see illustrations)**. Remove the bolts and label them; they are different lengths. Note the location of the side cover bracket.
15   Set the crankcase on a work surface, then separate the crankcase halves. **Note:** *Honda specifies setting the rear side down, then lifting the front half off the rear half. However, we found in our teardown of a 250 model that the rear half separated from the crankshaft, leaving the crankshaft and transmission shafts in the front half.* Don't pry against the mating surfaces or they'll develop leaks.
16   Remove the two crankcase dowels **(see illustration)**.
17   Refer to Sections 25 through 27 for information on the internal components of the crankcase.

Reassembly

18   Remove all traces of old gasket and sealant from the crankcase mating surfaces with a sharpening stone or similar tool. Be careful not

6   Remove the gearshift plate, stopper arm and cam, gearshift spindle and reverse stopper arm (see Section 24).
7   Remove the valve cover, rocker arms and pushrods, cylinder head, cylinder and piston (see Sections 7, 8, 9, 10, 13 and 14).
8   Remove the oil pump (see Section 19).

25.14a Here are the 350 rear crankcase bolts (arrows)

25.14b Here are the 250 rear crankcase bolts (arrows)

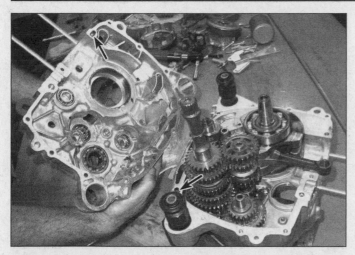

25.16 Separate the case halves and locate the dowels (arrows) (250 shown; 350 similar)

26.2a The 350 has two oil strainer screens (arrows)

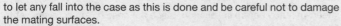

to let any fall into the case as this is done and be careful not to damage the mating surfaces.

19 Check to make sure the two dowel pins are in place in their holes in the mating surface of the rear crankcase half **(see illustration 25.16)**.

20 Reinstall the oil strainers and plate if they were removed.

21 Pour some engine oil over the transmission gears, the crankshaft bearing surface and the shift drum. Don't get any oil on the crankcase mating surface.

22 If you're working on a 350 model, apply a thin coat of sealant to the crankcase mating surface. If you're working on a 250, install a new crankcase gasket.

23 Carefully place the front crankcase half onto the rear crankcase half. While doing this, make sure the transmission shafts, shift drum, crankshaft and balancer fit into their ball bearings in the front crankcase half.

24 Install the rear crankcase half bolts in the correct holes (don't forget the side cover bracket) and tighten them so they are just snug. Then tighten them in two or three stages, in a criss-cross pattern, to the torque listed in this Chapter's Specifications.

25 Install the output shaft driven gear, then the drive gear. The splines inside the driven gear face away from then engine.

26 Turn the transmission mainshaft to make sure it turns freely. Also make sure the crankshaft turns freely.

27 The remainder of installation is the reverse of removal.

---

26 Crankcase components - inspection and servicing

Refer to illustrations 26.2a, 26.2b, 26.4, 26.5a and 26.5b

1 Separate the crankcase and remove the following:

 a) *Transmission shafts and gears*
 b) *Final drive shaft and gear*
 c) *Crankshaft and main bearings*
 d) *Balancer (350 models)*
 e) *Shift drum and forks*
 f) *Oil strainer screens and plate*

2 Clean the oil strainer screens thoroughly with solvent and check them for damage **(see illustrations)**. If they're damaged or can't be cleaned, replace them.

3 Clean the crankcase halves thoroughly with new solvent and dry them with compressed air. All oil passages should be blown out with compressed air and all traces of old gasket sealant should be removed from the mating surfaces. **Caution:** *Be very careful not to nick or gouge the crankcase mating surfaces or leaks will result. Check both crankcase sections very carefully for cracks and other damage.*

26.2b The 250 has one strainer screen, backed by a plate

4 If you're working on a 350 model, check the bearings in the case halves and in the crankcase front cover **(see illustration 26.2a and the accompanying illustration)**. If they don't turn smoothly, replace them. To remove bearings that aren't accessible from the outside, a blind hole puller will be needed (these can be rented). Drive the remaining

26.4 On 350 models, check the bearings inside the front cover

26.5a On 250 models, inspect the bearings in the front case half . . .

26.5b . . . and in the rear case half

27.2 Remove the washer (arrow) and sub-gearshift spindle

27.3 Lift the output shaft and remove its washer

bearings out with a bearing driver or a socket having an outside diameter slightly smaller than that of the bearing outer race. Before installing the bearings, allow them to sit in the freezer overnight, and about

fifteen-minutes before installation, place the case half in an oven, set to about 200-degrees F, and allow it to heat up. The bearings are an interference fit, and this will ease installation. **Warning:** *Before heating the case, wash it thoroughly with soap and water so no explosive fumes are present. Also, don't use a flame to heat the case.* Install the ball bearings with a socket or bearing driver that bears against the bearing outer race. Install the needle roller bearing with a shouldered drift that fits inside the bearing to keep it from collapsing while the shoulder applies pressure to the outer race.

5   If you're working on a 250 model, check the bearings in the front and rear case halves, as well as the countershaft oil seal **(see illustrations)**. Inspect the bearings, and replace them if necessary, as described in Step 4. It's a good idea to replace the countershaft oil seal whenever the crankcase is disassembled. Pry out the old seal, taking care not to damage the seal bore. Tap in a new one using a seal driver or socket the same diameter as the seal.

6   If any damage is found that can't be repaired, replace the crankcase halves as a set.

7   Assemble the case halves (see Section 25) and check to make sure the crankshaft and the transmission shafts turn freely.

27 Transmission shafts and shift drum - removal, inspection and installation

Note: *When disassembling the transmission shafts, place the parts on a long rod or thread a wire through them to keep them in order and facing the proper direction.*

Removal and disassembly

1   Remove the engine, then separate the case halves (see Sections 6 and 25).

350 models
Refer to illustrations 27.2, 27.3, 27.4a, 27.4b, 27.5, 27.6, 27.7a and 27.7b, 27.8a and 27.8b

2   Note the location of the washer on the sub-gearshift spindle, then lift it out of the rear case half **(see illustration)**.

3   Lift the output shaft and remove its washer **(see illustration)**.

4   Note how the shift forks engage the shift drum, then pull the shift fork shaft out of the forks **(see illustrations)**. Lift out the shift drum and remove the forks. As you remove the forks, look for labels on the upper side of each fork (F for front, C for center and RR for rear). If you don't see them, mark the upper side of each fork with a scribe or other sharp tool so they can be reinstalled in the correct locations.

27.4a Note how the shift forks engage the shift drum . . .

27.4b . . . then pull the shift fork shaft out of the forks

27.5 Remove the reverse idler gear and its shaft (arrow)

27.6 Lift the mainshaft and countershaft out of the transmission together

5   Remove the reverse idler gear and its shaft **(see illustration)**.
6   Lift the mainshaft and countershaft out of the transmission together **(see illustration)**.

7   Slide the components off the mainshaft and place them in order on a long rod or piece of wire (such as a straightened coat hanger) **(see illustrations)**.

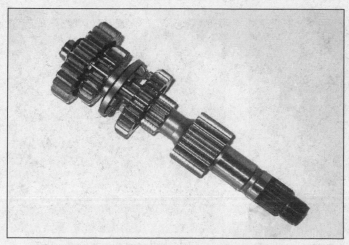

27.7a Here's the assembled 350 mainshaft

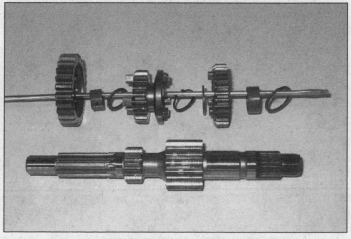

27.7b 350 mainshaft components

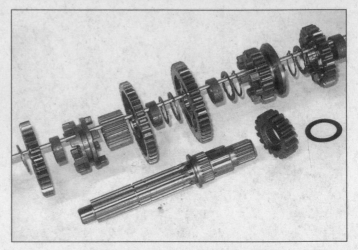

27.8a Here's the assembled 350 countershaft

27.8b 350 countershaft components

27.13a Note how the shift forks engage the shift drum and gears, then pull out the shaft

8   Slide the components off the countershaft, removing the snap-rings as you do so **(see illustration)**. Place them in order on a long rod or piece of wire (such as a straightened coat hanger) **(see illustration)**.

250 models

With gears in rear case half

Note: *This procedure applies if the front case half was lifted off, leaving the gears in the rear case half.*

9   Remove the countershaft thrust washer, first gear and bushing.

10   Remove the thrust washer, reverse idler gear, second thrust washer and reverse idler shaft.

11   Lift the mainshaft, countershaft, shift drum and forks out of the case as an assembly.

12   Note how the shift forks engage the shift drum and gears, then separate the shift forks and shaft from the shift drum. For the time being, leave the forks on the shaft.

With gears in front case half

Refer to illustrations 27.13b, 27.13c, 27.14, 27.15, 27.16, 27.17 and 27.18

Note: *This procedure applies if the rear case half was lifted off, leaving the gears in the front case half.*

13   Note how the shift forks engage the shift drum and gears, then pull out the shaft and separate the shift forks from the shift drum **(see illustration)**. Reassemble the forks on the shaft right away to ease installation **(see illustrations)**.

14   Take the thrust washer off the mainshaft **(see illustration)**.

15   Lift the shift drum out of the crankcase **(see illustration)**.

16   Lift the mainshaft and countershaft out of the crankcase together, noting that the reverse counter shifter and first gear may slide off the countershaft as you do so **(see illustration)**.

27.13b Reassemble the forks on the shaft right away to ease installation

27.13c The forks have different labels on each side

27.14 Take the thrust washer off the mainshaft

27.15 Lift the shift drum out of the crankcase

27.16 Lift the mainshaft and countershaft out of the crankcase together; the reverse counter shifter and first gear may slide off the countershaft

27.17 Remove the countershaft thrust washer from the front case half

27.18 Remove the reverse idler shaft and gear (and thrust washers on 2001 and earlier models)

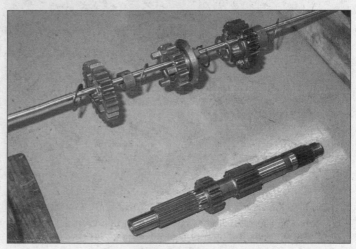

27.19 250 mainshaft components

17   Remove the countershaft thrust washer from the front case half **(see illustration)**.

18   Remove the reverse idler shaft and gear **(see illustration)**. **Note:** *On 2001 and earlier models, there's a thrust washer on each side of the gear.*

All 250 models

Refer to illustrations 27.19 and 27.20

19   Slide the components off the mainshaft, removing the snap-ring as you do so, and place them in order on a long rod or piece of wire (such as a straightened coat hanger) **(see illustration)**.

20   Slide the components off the countershaft, removing the snap-rings as you do so. Place them in order on a long rod or piece of wire (such as a straightened coat hanger) **(see illustration)**.

Inspection

Refer to illustration 27.23

21   Wash all of the components in clean solvent and dry them off.

22   Inspect the shift fork grooves in the countershaft second/reverse shifter, mainshaft third gear and countershaft fourth gear. If a groove is worn or scored, replace the affected part and inspect its corresponding shift fork.

27.20 250 countershaft components

27.23 Check the fork fingers and pins and the shift drum grooves for wear

27.30a Align the oil holes in bushings and shafts

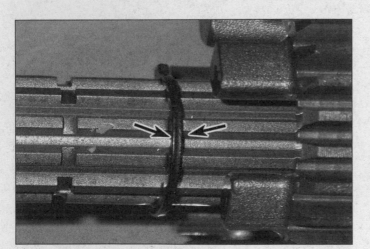

27.30b The rounded side of each snap-ring and thrust washer (right arrow) faces the direction of thrust; the flat side (left arrow) faces away

pins for excessive wear and distortion and replace any defective parts with new ones.

24   Measure the inside diameter of the forks and the outside diameter of the fork shaft and compare to the values listed in this Chapter's Specifications. Replace any parts that are worn beyond the limits. Check the shift fork shaft for evidence of wear, galling and other damage. Make sure the shift forks move smoothly on the shafts. If the shafts are worn or bent, replace them with new ones.

25   Check the edges of the grooves in the shift drum for signs of excessive wear.

26   Spin the shift drum bearing with fingers. Replace the bearing if it's rough, loose or noisy.

27   Check the gear teeth for cracking and other obvious damage. Check the bushing and surface in the inner diameter of the freewheeling gears for scoring or heat discoloration. Replace damaged parts.

28   Inspect the engagement dogs and dog holes on gears so equipped for excessive wear or rounding off. Replace the paired gears as a set if necessary.

29   Check the mainshaft needle bearing in the crankcase for wear or heat discoloration and replace them if necessary (see Section 26).

Assembly

Refer to illustrations 27.30a, 27.30b, 27.30c and 27.30d

30   Assembly is the basically the reverse of the disassembly procedure, but take note of the following points:

  a) *Oil holes in bushings must be aligned with the corresponding oil hole in the shaft the bushing is installed on* **(see illustration)**.

23   Check the shift forks for distortion and wear, especially at the fork fingers **(see illustration)**. Measure the thickness of the fork fingers and compare your findings with this Chapter's Specifications. If they are discolored or severely worn they are probably bent. Inspect the guide

27.30c Position each snap-ring so its ends are supported by a raised spline

27.30d Make sure the gears mesh properly before installing them in the case

28.2a Press the crankshaft out of the rear crankcase half . . .

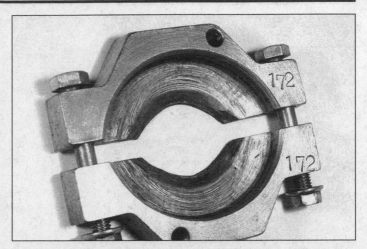

28.2b . . . if the bearing stays on the crankshaft, remove it with the press and a bearing splitter

b) *Thrust washers and snap-rings have a rounded edge and a flat edge. The rounded edge faces toward the direction of thrust (that is, toward the gear whose thrust is being controlled)* **(see illustration)**.

c) *Snap-rings should be installed so that the snap-ring gap aligns with a gap in the shaft splines, so that a raised spline gives support to each end of the snap-ring* **(see illustration)**.

d) *Always use new snap-rings.*

e) *Lubricate the components with engine oil before assembling them.*

f) *On 250 models, don't install countershaft first gear or its bushing yet. They will be installed later.*

g) *After assembling the countershaft and mainshaft, check the gears to make sure they mesh correctly before installing the shafts in the crankcase* **(see illustration)**.

Installation

350 models

31 Mesh the mainshaft and countershaft and install them together into the crankcase.

32 Place the reverse idler shaft in the crankcase, then install the gear.

33 Mesh the shift fork labeled RR with the groove on mainshaft third gear (label upward, toward the front crankcase half).

34 Mesh the center fork labeled C with countershaft fourth gear (label upward, toward the front crankcase half).

35 Mesh the front fork labeled F with the gear shifter (label upward, toward the front crankcase half).

36 Lubricate the shift drum with clean engine oil and install it, meshing the fork guide pins with the shift drum grooves.

37 Lubricate the fork shaft with clean engine oil and install it in the forks. Make sure it passes through all three forks and seats securely in the crankcase.

38 Install the output shaft and its washer.

39 Install the sub-gearshift spindle and its washer.

40 The remainder of installation is the reverse of the removal steps.

250 models

Note: *The following steps describe assembling the transmission into the rear crankcase half.*

41 Assemble the shift forks on the shaft in the correct order.

42 Mesh the shift fork labeled HB3L (1997 through 2004) or 8KR (2005) with the groove on mainshaft third gear (label upward, toward the front crankcase half).

43 Mesh the center fork labeled HB3C (1997 through 2004) or 8KC (2005) with countershaft fourth gear (label upward, toward the front crankcase half).

44 Mesh the front fork labeled HB3R (1997 through 2004) or HNGF (2005) with the countershaft gear shifter (label upward, toward the front crankcase half).

45 Engage the shift fork pins with the grooves in the shift drum. Mesh the mainshaft and countershaft and install them, together with the shift

drum and forks, into the crankcase.

46 Place the reverse idler shaft in the crankcase, then install the thrust washer (1997 through 2001 only), gear (all models) and second thrust washer(1997 through 2001 only).

47 Install countershaft first gear, its bushing and thrust washer.

48 The remainder of installation is the reverse of the removal steps.

28 Crankshaft and balancer - removal, inspection and installation

Note: *The procedures in this section require a press and special tools. If you don't have the necessary equipment or suitable substitutes, have the crankshaft removed and installed by a Honda dealer.*

Removal

Refer to illustrations 28.2a and 28.2b

1 Remove the engine, separate the crankcase halves and remove the transmission (Sections 6, 25 and 27).

2 Place the rear crankcase half in a press and press out the crankshaft, removing the balancer at the same time on 350 models **(see illustration)**. The ball bearing may remain in the crankcase or come out with the crankshaft. If it stays on the crankshaft, remove it with the press and a bearing splitter **(see illustration)**. Discard the bearing, no matter what its apparent condition, and use a new one on installation.

Inspection

Refer to illustrations 28.3, 28.4, 28.6 and 28.7

3 Measure the side clearance between connecting rod and crankshaft with a feeler gauge **(see illustration)**. If it's more than the limit

28.3 Check the connecting rod side clearance with a feeler gauge

28.4 Check the connecting rod radial clearance with a dial indicator

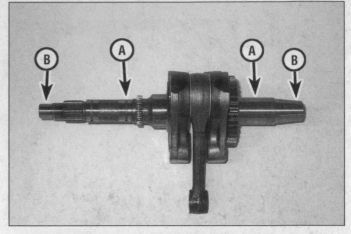

28.6 Place a V-block on each side of the crankshaft (A) and measure runout 6 mm in from the ends (B)

listed in this Chapter's Specifications, replace the crankshaft and connecting rod as an assembly.

4   Set up the crankshaft in V-blocks with a dial indicator contacting the big end of the connecting rod **(see illustration)**. Move the connecting rod up-and-down against the indicator pointer and compare the reading to the value listed in this Chapter's Specifications. If it's beyond the limit, replace the crankshaft and connecting rod as an assembly.

5   Check the crankshaft gear, sprockets and bearing journals for visible wear or damage, such as chipped teeth or scoring. If any of these conditions are found, replace the crankshaft and connecting rod as an assembly.

6   Set the crankshaft in a pair of V-blocks, with a dial indicator contacting each end **(see illustration)**. Rotate the crankshaft and note the runout. If the runout at either end is beyond the limit listed in this Chapter's Specifications, replace the crankshaft and connecting rod as an assembly.

7   If you're working on a 350 model, check the balancer gear and bearing journals for visible wear or damage and replace the balancer if any problems are found **(see illustration)**.

Installation

Refer to illustrations 28.13, 28.14, 28.15, 28.17a and 28.17b

8   Installation requires a special puller and adapter or their equivalents. Read through the procedure before starting; if you don't have access to the necessary tools, have the crankshaft installed by a dealer service department or machine shop.

9   In the US, the tools are:
a) *Assembly collar 07965-VM00100*
b) *Assembly shaft 07931-ME4010B*
c) *Special nut 07931-HB3020A*
d) *Threaded adapter 07931-KF00200*

10   In all markets except the US, the tools are:
a) *Tool set 07965-VM00000*
b) *Assembly collar 07965-VM00100*
c) *Assembly shaft 07965-VM00200*
d) *Threaded adapter 07965-VM00300*

11   Install the balancer bearing (350 models) and crankshaft bearing in the rear crankcase half (see Section 26).

12   If you're working on a 350 model, install the crankshaft and balancer in the rear crankcase half. The balancer drive gear on the crankshaft and the balancer driven gear on the balancer have lines stamped in their outer edges. These are timing marks, which must be aligned with each other when the crankshaft and balancer are installed. **Note:** *It's important to align the timing marks exactly throughout this procedure. Severe engine vibration will occur if the crankshaft and balancer are out of time.*

13   Thread a puller adapter into the end of the crankshaft **(see illustration)**.

14   Temporarily install the front crankcase half on the rear crankcase **(see illustration)**.

15   Install a crankshaft puller over the crankshaft and install the special nut **(see illustration)**.

28.7 Check the balancer gear and bearing journals for wear or damage

28.13 Thread the special tool adapter into the end of the crankshaft . . .

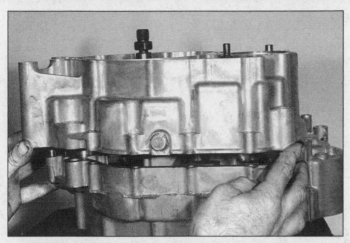

28.14 . . . install the front crankcase half without bolts . . .

28.15 . . . and install the puller on the crankshaft adapter and crankcase half; hold the shaft (upper arrow) with a wrench and turn the nut to pull the crankshaft into the bearing

16 Hold the puller shaft with one wrench and turn the nut with another wrench to pull the crankshaft into the center race of the ball bearing.

17 Remove the special tools. Lift the rear crankcase half off and check to make sure the timing marks on the other side of the crankshaft and balancer are aligned (see illustrations).

18 The remainder of installation is the reverse of the removal steps.

29 Initial start-up after overhaul

1 Make sure the engine oil level is correct, then remove the spark plug from the engine. Place the engine kill switch in the Off position and unplug the primary (low tension) wires from the coil.

2 Turn on the key switch and crank the engine over with the starter several times to build up oil pressure. Reinstall the spark plug, connect the wires and turn the switch to On.

3 Make sure there is fuel in the tank, then operate the choke.

4 Start the engine and allow it to run at a moderately fast idle until it reaches operating temperature. **Caution:** *If the oil temperature light doesn't go off, or it comes on while the engine is running, stop the engine immediately.*

5 Check carefully for oil leaks and make sure the transmission and controls, especially the brakes, function properly before road testing the machine. Refer to Section 29 for the recommended break-in procedure.

6 Upon completion of the road test, and after the engine has cooled down completely, recheck the valve clearances (see Chapter 1).

30 Recommended break-in procedure

1 Any rebuilt engine needs time to break-in, even if parts have been installed in their original locations. For this reason, treat the machine gently for the first few miles to make sure oil has circulated throughout the engine and any new parts installed have started to seat.

2 Even greater care is necessary if the cylinder has been rebored or a new crankshaft has been installed. In the case of a rebore, the engine will have to be broken in as if the machine were new. This means greater use of the transmission and a restraining hand on the throttle for the first few operating days. There's no point in keeping to any set speed limit - the main idea is to vary the engine speed, keep from lugging (laboring) the engine and to avoid full-throttle operation. These recommendations can be lessened to an extent when only a new crankshaft is installed. Experience is the best guide, since it's easy to tell when an engine is running freely.

28.17a Make sure the balancer timing mark on the crankshaft weight (left arrow) is aligned with the slot in the balancer shaft (right arrow) . . .

28.17b . . . and the groove on the balance gear (lower arrow) aligns with the mark on the crankshaft (upper arrow)

3 If a lubrication failure is suspected, stop the engine immediately and try to find the cause. If an engine is run without oil, even for a short period of time, irreparable damage will occur.

Notes

Notes

Notes

Chapter 3
Fuel and exhaust systems

Contents

	Section		Section
Air cleaner housing - removal and installation	9	Exhaust system - removal and installation	12
Carburetor - reassembly and float height check	8	Fuel tank - cleaning and repair	3
Carburetor - removal and installation	6	Fuel tank - removal and installation	2
Carburetor heater (350 models) - removal and installation	13	General information	1
Carburetor overhaul - general information	5	Idle fuel/air mixture adjustment	4
Carburetors - disassembly, cleaning and inspection	7	Throttle cable and housing - removal, installation	
Choke cable - removal and installation	11	and adjustment	10

Specifications

General
Fuel type .... Unleaded or low lead gasoline (petrol) subject to local regulations; minimum octane 91 RON (87 pump octane)

350 models
Carburetor
Identification mark VE94A
Jet sizes and settings
Main jet
2000 through 2003 models
Standard 130
High altitude 125
2004 and later models
Standard 128
High altitude 122
Jet needle clip position .... Third groove from top
Slow jet
2000 through 2003 models 42
2004 and later models 42
Pilot screw
Sea level to 5000 feet, turns out from lightly seated position 1-3/4
High altitude, turns in from sea level setting 3/4
Float level .... 18.5 mm (0.73 inch)
Heater resistance .... 13 to 15 ohms at 20-degrees C/68-degrees F
Torque settings
Carburetor intake tube clamp screw .... 4 Nm (35 inch-lbs)
Muffler clamp bolt .... 23 Nm (17 ft-lbs)
Exhaust heat shield bolts .... 22 Nm (16 ft-lbs)

250 models
Carburetor
Identification mark
Recon models
1997 except California PDC1B
1997 California, all 1998 through 2001 .... PDC1C
2002 through 2004 .... PDC1E
2005 and 2006 .... PDC1F
2007 and later PDC1H
TRX250EX models
2005 and earlier PDC1D
2006 and later PDC1G

250 models (continued)

Jet sizes and settings
  Main jet
    Standard 95
    High altitude 92
  Jet needle clip position
    1997 Recon, 2005 and earlier TRX250EX .... 3rd groove from top
    1998 through 2004 Recon .... 2nd groove from top
    2005 and later Recon, 2006 and later TRX250EX .... Not specified
  Slow jet
    All Recon, 2005 and earlier TRX250EX 38
    2006 and later TRX250EX 42
Pilot screw initial setting (sea level to 5,000 feet, turns out from lightly seated position)
  Recon models
    1997 except California 2-7/8
    1997 California 2-3/4
    1998 through 2001 .... 2-5/8
    2002 through 2004 2
    2005 and later 1-1/2
  TRX250EX models
    2005 and earlier 1-7/8
    2006 and later 1-1/2
Pilot screw final setting (sea level to 5000 feet, turns out from initial setting)
  Recon models
    1997 except California 3/4
    1997 California, all 1998 through 2001 models 7/8
    2002 through 2004 .... 1/2
    2005 and later 5/8
  TRX250EX models
    2005 and earlier .... Same as initial setting
    2006 and later 5/8
Pilot screw high altitude setting (3000 to 8000 feet, turns in from final sea level setting)
  All except 2006 and later Recon 1/8
  2006 and later Recon 1/4
Float level .... 14 mm (0.6 inch)

Torque settings

Carburetor insulator stud .... 10 Nm (84 inch-lbs)
Muffler clamp bolt .... 23 Nm (17 ft-lbs)
Exhaust heat shield bolts .... 22 Nm (16 ft-lbs)

1 General information

The fuel system consists of the fuel tank, fuel tap, filter screen, carburetor and connecting lines, hoses and control cables.

The carburetor used on 350 models is a constant vacuum unit with a butterfly-type throttle valve and an enrichment circuit for cold starting.

The carburetor used on 250 models is a piston-valve unit with a butterfly-type choke valve for cold starting.

The exhaust system consists of a pipe and muffler/silencer with a spark arrester function.

Many of the fuel system service procedures are considered routine maintenance items and for that reason are included in Chapter 1.

2 Fuel tank - removal and installation

Warning: *Gasoline is extremely flammable, so take extra precautions when you work on any part of the fuel system. Don't smoke or allow open flames or bare light bulbs near the work area, and don't work in a garage where a gas-type appliance (such as a water heater or clothes dryer) is present. Since gasoline is carcinogenic, wear protective gloves when there is a possibility of being exposed to fuel, and if you spill any fuel on your skin, rinse it off immediately with soap and water. Mop up any spills immediately and do not store fuel-soaked rags where they could ignite. When you perform any kind of work on the fuel system, wear safety glasses and have an extinguisher suitable for a class B type fire (flammable liquids) on hand.*

Removal

Refer to illustrations 2.3, 2.4, 2.5, 2.6a and 2.6b

1   Turn the fuel tap to Off.

2   If you're working on a 350 model, remove the seat, left and right side covers and the fuel tank cover. If you're working on a 250 model, remove the front fenders (see Chapter 7).

3   Remove the mounting bolts at the front of the fuel tank **(see illustration)**.

4   Remove the tank seal and unhook the mounting band on each side at the rear of the tank **(see illustration)**.

5   Lift the tank and disconnect the fuel line from the fuel tap **(see illustration)**. Lift the tank off the vehicle together with the fuel tap.

6   If necessary, remove the fuel tank heat shield, as well as the air guide plate on 350 models **(see illustrations)**.

Installation

7   Before installing the tank, check the condition of the tank mounts, seal and rubber mounting bands - if they're hardened, cracked, or show any other signs of deterioration, replace them **(see illustration 2.4)**.

8   When installing the tank, reverse the removal procedure. Do not pinch any control cables or wires.

2.3 Remove the mounting bolts at the front of the fuel tank (arrows)

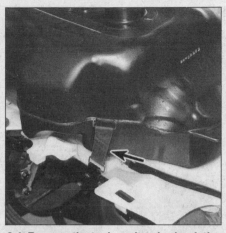

2.4 Remove the tank seal and unhook the mounting band on each side (arrow)

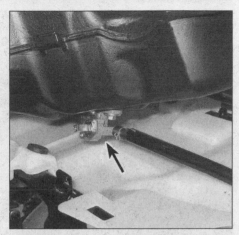

2.5 Lift the tank and disconnect the fuel line from the tap (arrow)

2.6a Here's the fuel tank heat shield on a Rancher (other models similar) . . .

2.6b . . . and the air guide plate

4.3a Here's the Rancher pilot screw (arrow) . . .

3 Fuel tank - cleaning and repair

1   All repairs to the fuel tank should be carried out by a professional who has experience in this critical and potentially dangerous work. Even after cleaning and flushing of the fuel system, explosive fumes can remain and ignite during repair of the tank.

2   If the fuel tank is removed from the vehicle, it should not be placed in an area where sparks or open flames could ignite the fumes coming out of the tank. Be especially careful inside garages where a natural gas-type appliance is located, because the pilot light could cause an explosion.

4 Idle fuel/air mixture adjustment

Normal adjustment

Refer to illustrations 4.3a and 4.3b

1   Idle fuel/air mixture on these vehicles is preset at the factory and should not need adjustment unless the carburetor is overhauled or the pilot screw, which controls the mixture adjustment, is replaced.

2   The engine must be properly tuned up before making the adjustment (valve clearances set to specifications, spark plug in good condition and properly gapped).

3   To make an initial adjustment, turn the pilot screw clockwise until it seats lightly, then back it out the number of turns listed in this Chapter's Specifications **(see illustrations)**. Caution: *Turn the screw just far enough to seat it lightly. If it's bottomed hard, the screw or its seat may*

be damaged, which will make accurate mixture adjustments impossible.

4   Warm up the engine to normal operating temperature. Shut it off and connect a tune-up tachometer, following the tachometer manufacturer's instructions.

5   Restart the engine and compare idle speed to the value listed in the Chapter 1 Specifications. Adjust it if necessary.

6   Turn the pilot screw a little at a time, in or out, to obtain the highest possible idle speed. Once you've done this, reset the idle to the specified idle speed using the throttle stop screw.

4.3b . . . and here's the Recon/TRX250EX pilot screw (arrow)

6.4 On Rancher models, disconnect the heater connector

6.5 Loosen the clamping band on the intake manifold (A); disconnect the vent hose (B) and starting enrichment valve (C) - on installation, align the notch and tab (D)

7   Turn the pilot screw in (clockwise) slowly until engine speed drops 100 rpm below the specified idle speed.

8   If you're working on a 350 model, back the pilot screw out (counterclockwise) to the final setting listed in this Chapter's Specifications.

9   If you're working on a Recon model, back the pilot screw out (counterclockwise) the final fraction of a turn listed in this Chapter's Specifications.

High altitude adjustment

10   If the vehicle is normally used at altitudes from sea level to 5000 feet (1500 meters), use the normal main jet and pilot screw setting. If it's used at altitudes between 3000 and 8000 feet (1000 and 2500 meters), the main jet and pilot screw setting must be changed to compensate for the thinner air. **Caution:** *Don't use the vehicle for sustained operation below 5000 feet (15 meters) with the main jet and pilot screw at the high altitude settings or the engine may overheat and be damaged.*

11   Refer to Section 7 and change the main jet to the high altitude jet listed in this Chapter's Specifications.

12   Set the pilot screw to the standard setting, then turn it in the additional amount listed in this Chapter's Specifications.

13   With the vehicle at high altitude, refer to Chapter 1 and adjust the idle speed.

---

5   Carburetor overhaul - general information

1   Poor engine performance, hesitation, hard starting, stalling, flooding and backfiring are all signs that major carburetor maintenance may be required.

2   Keep in mind that many so-called carburetor problems are really not carburetor problems at all, but mechanical problems within the engine or ignition system malfunctions. Try to establish for certain that the carburetor is in need of maintenance before beginning a major overhaul.

3   Check the fuel tap and its strainer screen, the in-tank fuel strainer, the fuel lines, the intake manifold clamps, the O-ring between the intake manifold and cylinder head, the vacuum hoses, the air filter element, the cylinder compression, the spark plug and the ignition timing before assuming that a carburetor overhaul is required. If the vehicle has been unused for more than a month, refer to Chapter 1, drain the float chamber and refill the tank with fresh fuel.

4   Most carburetor problems are caused by dirt particles, varnish and other deposits which build up in and block the fuel and air passages. Also, in time, gaskets and O-rings shrink or deteriorate and cause fuel and air leaks which lead to poor performance.

5   When the carburetor is overhauled, it is generally disassembled completely and the parts are cleaned thoroughly with a carburetor cleaning solvent and dried with filtered, unlubricated compressed air.

The fuel and air passages are also blown through with compressed air to force out any dirt that may have been loosened but not removed by the solvent. Once the cleaning process is complete, the carburetor is reassembled using new gaskets, O-rings and, generally, a new inlet needle valve and seat.

6   Before disassembling the carburetor, make sure you have a carburetor rebuild kit (which will include all necessary O-rings and other parts), some carburetor cleaner, a supply of rags, some means of blowing out the carburetor passages and a clean place to work.

---

6   Carburetor - removal and installation

Warning: *Gasoline is extremely flammable, so take extra precautions when you work on any part of the fuel system. Don't smoke or allow open flames or bare light bulbs near the work area, and don't work in a garage where a gas-type appliance (such as a water heater or clothes dryer) is present. Since gasoline is carcinogenic, wear protective gloves when there is a possibility of being exposed to fuel, and if you spill any fuel on your skin, rinse it off immediately with soap and water. Mop up any spills immediately and do not store fuel-soaked rags where they could ignite. When you perform any kind of work on the fuel system, wear safety glasses and have an extinguisher suitable for a class B type fire (flammable liquids) on hand.*

Removal

1   Remove the seat (see Chapter 7).

350 models

Refer to illustrations 6.4, 6.5, 6.8a and 6.8b

2   Remove the side covers (see Chapter 7).

3   Remove the air cleaner housing (see Section 9).

4   Free the carburetor heater wiring harness from its retainer and disconnect the connector **(see illustration)**.

5   Loosen the clamping band on the intake manifold tube **(see illustration)**. Work the carburetor free of the manifold and lift it up.

6   Disconnect the vent hose from the carburetor **(see illustration 6.5)**. Unscrew the starting enrichment valve nut and pull the valve out of the carburetor body. Follow the fuel line from the tap to the carburetor and disconnect it.

7   Refer to Section 10 and disconnect the throttle cable from the carburetor body. The carburetor can now be removed.

8   Check the intake manifold tube for cracks, deterioration or other damage. If it has visible defects, or if there's reason to suspect its O-ring is leaking, remove it and inspect the O-ring **(see illustrations)**.

9   After the carburetor has been removed, stuff clean rags into the

6.8a Remove the bolts (arrows) . . .

6.8b . . . then detach the manifold tube
and inspect its O-ring (arrow)

6.11a Unscrew the carburetor top . . .

6.11b . . . lift off the top and spring . . .

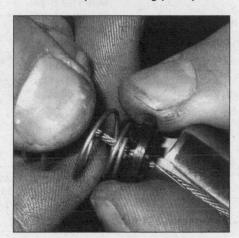

6.11c . . . and top gasket . . .

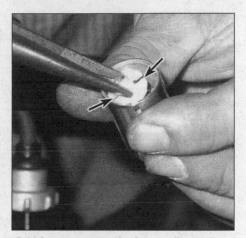

6.11d . . . squeeze the jet needle retainer
(note the position of the slots (arrows)) . . .

intake tube (or the intake port in the cylinder head, if the tube has been removed) to prevent the entry of dirt or other objects.

250 models

Refer to illustrations 6.11a through 6.11e, 6.12, 6.14, 6.16a and 6.16b

10   Remove the seat (see Chapter 7) and air cleaner housing (see Section 9).

11   Unscrew the carburetor top and pull it out, together with the jet needle **(see illustrations)**.

12   Detach the throttle piston and jet needle from the cable **(see illustration)**.

13   Make sure the float chamber drain hose is in place, then loosen the drain screw and drain the gasoline from the float bowl **(see illustration 17.9b in Chapter 1)**.

14   Detach the vent hose, fuel line and air cutoff valve hose (2005 models) from the carburetor **(see illustration)**.

15   Disconnect the choke cable from the lever (see Section 11).

6.11e . . . and remove the retainer and
spring; on assembly, align the dimple with
the round depression (arrows)

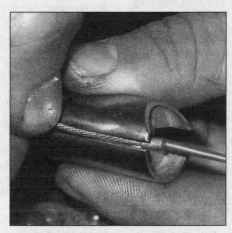

6.12 Compress the spring and slip and
cable out of the piston slot

6.14 Disconnect the vent hose, fuel line
and air cutoff valve hose (2005)

6.16a Unscrew the mounting nuts and take the carburetor off . . .

6.16b . . . and remove the O-ring

16   Unscrew the carburetor mounting nuts **(see illustration)**. Take the carburetor off the studs and remove the O-ring **(see illustration)**.

Installation

Refer to illustration 6.17

17   Installation is the reverse of the removal Steps, with the following additions:

a) *On 350 models, use a new O-ring if the manifold was removed from the cylinder head. On 250 models, use a new O-ring whenever the carburetor is removed.*

b) *On 350 models, lubricate the threads of the starting enrichment valve with multipurpose grease before installing it **(see illustration)**.*

c) *On 350 models, align the tab on the carburetor with the notch in the intake manifold.*

d) *Adjust the throttle freeplay and idle speed (see Chapter 1).*

7   Carburetors - disassembly, cleaning and inspection

Warning: *Gasoline is extremely flammable, so take extra precautions when you work on any part of the fuel system. Don't smoke or allow open flames or bare light bulbs near the work area, and don't work in a garage where a gas-type appliance (such as a water heater or clothes dryer) is present. Since gasoline is carcinogenic, wear protective gloves when there is a possibility of being exposed to fuel, and if you spill any fuel on your skin, rinse it off immediately with soap and water. Mop up any spills immediately and do not store fuel-soaked rags where they*

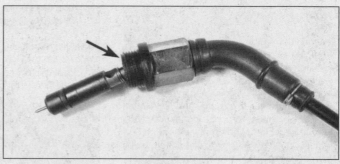

6.17 Check the starting enrichment valve for wear and damage and replace it if necessary; apply a thin coat of multipurpose grease to the end of the valve threads (arrow) on installation

could ignite. When you perform any kind of work on the fuel system, wear safety glasses and have an extinguisher suitable for a class B type fire (flammable liquids) on hand.

Disassembly

350 models

Refer to illustrations 7.2a through 7.2t

1   Remove the carburetor from the machine as described in Section 6. Set it on a clean working surface. Remove the carburetor heater (see Section 13).

2   Refer to the accompanying illustrations to disassemble the carburetor **(see illustrations)**.

7.2a Remove the air cutoff valve, its jet and O-rings (left arrow) and the passage O-ring (right arrow)

7.2b Remove the four screws securing the vacuum chamber cover to the carburetor body (arrows)

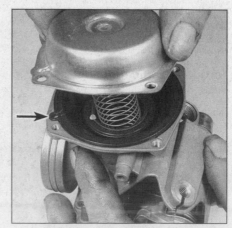

7.2c Lift the cover off and remove the piston spring; note the location of the tab (arrow) which fits into a notch

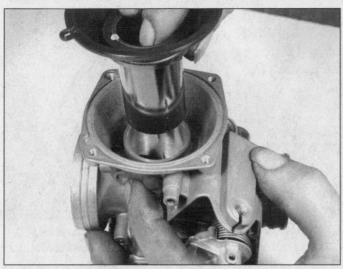

7.2d Peel the diaphragm away from its groove in the carburetor body, being careful not to tear it, and lift out the diaphragm/piston assembly

7.2e Push in on the needle jet holder and turn it 90-degrees counterclockwise with an 8 mm socket . . .

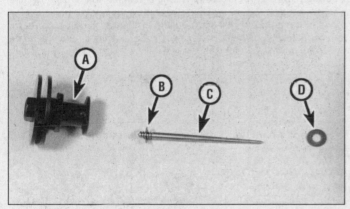

7.2f . . . then remove the holder and separate the needle, clip and washer from the piston; standard position for the clip is in the third groove

a   *Holder*   c   *Jet needle*
b   *Clip*   d   *Washer*

7.2g Remove the four screws retaining the float chamber to the carburetor body (arrows) . . .

7.2h . . . then detach the float chamber and remove its O-ring

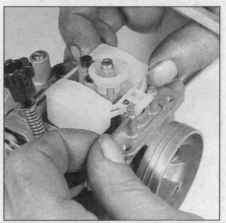

7.2i Remove the float chamber baffle, push the float pivot pin out and detach the float (and fuel inlet valve needle) from the carburetor body

7.2j Turn the pilot screw in, counting the number of turns until it bottoms lightly, and record the number for use when installing the screw . . .

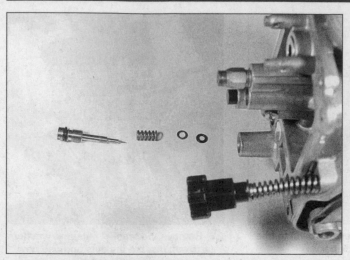

7.2k . . . then remove the pilot screw along with its spring, washer and O-ring

7.2l Unscrew the starter jet

250 models

Refer to illustrations 7.4a through 7.4n

3   Remove the carburetor from the machine as described in Section 6. Set it on a clean working surface.

4   Refer to the accompanying illustrations to disassemble the carburetor **(see illustrations)**.

Cleaning

Caution: *Use only a carburetor cleaning solution that is safe for use with plastic parts (be sure to read the label on the container).*

5   Submerge the metal components in the carburetor cleaner for approximately thirty minutes (or longer, if the directions recommend it).

6   After the carburetor has soaked long enough for the cleaner to

7.2m Unscrew the slow jet and pull it out . . .

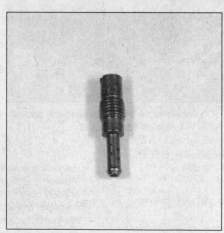

7.2n . . . noting that the narrow end goes in first

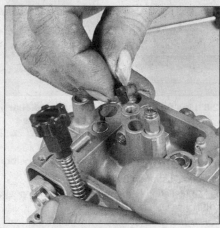

7.2o Remove the rubber plug from its passage

7.2p Unscrew the main jet from the needle jet holder . . .

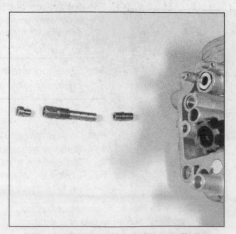

7.2q . . . then unscrew the needle jet holder and remove the needle jet

7.2r Remove the float chamber drain screw and its O-ring

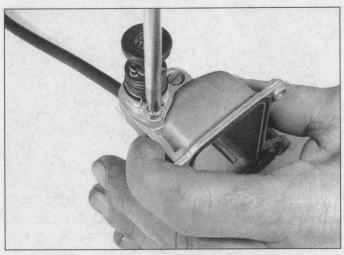

7.2s Remove the primer knob mounting screws . . .

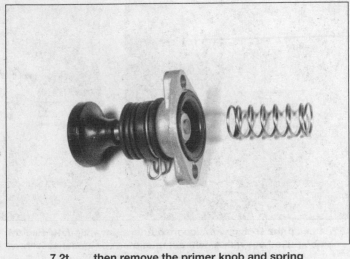

7.2t . . . then remove the primer knob and spring

loosen and dissolve most of the varnish and other deposits, use a brush to remove the stubborn deposits. Rinse it again, then dry it with compressed air. Blow out all of the fuel and air passages in the main and upper body. **Caution:** *Never clean the jets or passages with a piece of wire or a drill bit, as they will be enlarged, causing the fuel and air metering rates to be upset.*

Inspection

7   Check the tapered portion of the pilot screw for wear or damage. Replace the pilot screw if necessary.

8   Check the carburetor body, float chamber and vacuum chamber cover for cracks, distorted sealing surfaces and other damage. If any

7.4a On 2005 Recon models, disconnect the air cutoff valve hose . . .

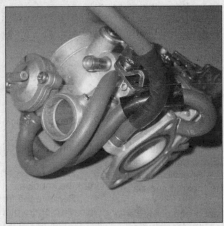

7.4b . . . note the routing of the other hoses and remove them . . .

7.4c . . . remove the air cutoff valve cover . . .

7.4d . . . and diaphragm . . .

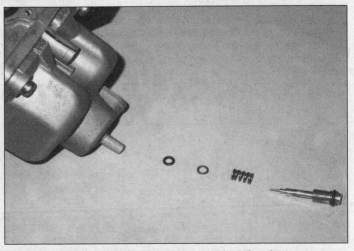

7.4e . . . Drill out the cover and remove the idle mixture screw, spring, washer and O-ring . . .

7.4f . . . if the O-ring won't come out, it can be pulled out with a bent paper clip

7.4g Remove the float chamber screws (arrows) . . .

7.4h . . . take off the float chamber and remove its O-ring . . .

defects are found, replace the faulty component, although replacement of the entire carburetor will probably be necessary (check with your parts supplier for the availability of separate components).

9   Check the jet needle for straightness by rolling it on a flat surface (such as a piece of glass). Replace it if it's bent or if the tip is worn.

10   Check the tip of the fuel inlet valve needle. If it has grooves or scratches in it, it must be replaced. Push in on the rod in the other end

of the needle, then release it - if it doesn't spring back, replace the valve needle.

11   Check the O-rings on the float chamber and the drain plug (in the float chamber). Replace them if they're damaged.

12   Operate the throttle shaft (350 models) or choke lever (250 models) to make sure the throttle or choke butterfly valve opens and closes smoothly. If it doesn't, replace the carburetor.

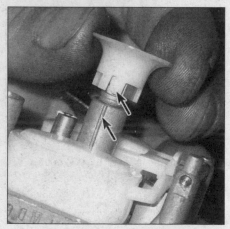

7.4i . . . lift off the baffle, noting how its notch aligns with the raised rib (arrows) . . .

7.4j . . . hold the needle jet holder with a wrench and unscrew the main jet, then unscrew the needle jet holder and remove the needle jet . . .

7.4k . . . unscrew the slow jet . . .

7.4l . . . tap out the float pivot pin with a punch, noting that one end is swaged . . .

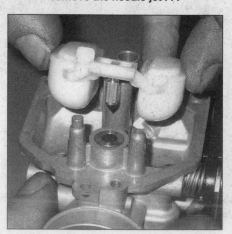

7.4m . . . lift off the float together with the needle valve . . .

7.4n . . . and detach the needle valve from the float

9.2 Disconnect the air vent tubes (arrow) from the intake duct

9.3 Disconnect the crankcase breather hose from the underside of the housing (arrow)

13   Check the floats for damage. This will usually be apparent by the presence of fuel inside one of the floats. If the floats are damaged, they must be replaced.

14   If you're working on a 350 model, check the diaphragm for splits, holes and general deterioration. Holding it up to a light will help to reveal problems of this nature.

15   Insert the piston in the carburetor body and see that it moves up-and-down smoothly. Check the surface of the piston for wear. If it's worn excessively or doesn't move smoothly in the bore, replace the carburetor.

16   If you're working on a 350 model, check the starting enrichment valve for wear or damage and replace it if any defects are found **(see illustration 6.17)**.

---

8   Carburetor - reassembly and float height check

Caution: *When installing the jets, be careful not to over-tighten them - they're made of soft material and can strip or shear easily.*

1   Install the clip on the jet needle if it was removed. Place it in the needle groove listed in this Chapter's Specifications. Install the washer, needle and clip in the piston, then install the retainer. If you're working on a 350 model, the retainer down and turn it 90-degrees clockwise with an 8 mm socket.

2   Install the pilot screw (if removed) along with its spring, washer and O-ring, turning it in until it seats lightly. Now, turn the screw out the number of turns that was previously recorded.

3   If you're working on a 350 model, install the diaphragm/vacuum piston assembly into the carburetor body. Lower the spring into the piston. Seat the bead of the diaphragm into the groove in the top of the carburetor body, making sure the diaphragm isn't distorted or kinked **(see illustration 7.2c)**. This is not always an easy task. If the diaphragm seems too large in diameter and doesn't want to seat in the groove, place the vacuum chamber cover over the carburetor diaphragm, insert your finger into the throat of the carburetor and push up on the vacuum piston, holding it almost all the way up. Push down gently on the vacuum chamber cover - it should drop into place, indicating the diaphragm has seated in its groove. Once this occurs, install at least two of the vacuum chamber cover screws before you let go of the piston. Install the remaining vacuum chamber cover screws and tighten all of them securely.

4   Reverse the disassembly steps to install the jets and other parts.

5   Invert the carburetor. Attach the fuel inlet valve needle to the float. Set the float into position in the carburetor, making sure the valve needle seats correctly. Install the float pivot pin. To check the float height, hold the carburetor so the float hangs down, then tilt it back until the valve needle is just seated. Measure the distance from the float cham-

ber gasket surface to the float surface farthest from the gasket surface. Compare your measurement to the float level listed in this Chapter's Specifications. There's no means of adjustment; if it isn't as specified, replace the float and needle valve.

6   Install the O-ring into the groove in the float chamber. Place the float chamber on the carburetor and install the screws, tightening them securely.

7   If you're working on a 350 model, install the starting enrichment valve.

---

9   Air cleaner housing - removal and installation

Removal
350 models
Refer to illustrations 9.2, 9.3, 9.4 and 9.5

1   Remove the seat, left side cover and fuel tank cover (see Chapter 7).

2   Locate the air intake duct to the left of the fuel tank **(see illustration)**. Label the air vent tubes, then disconnect them from the duct.

3   Disconnect the crankcase breather hose from the underside of the housing **(see illustration)**.

4   Loosen the connecting tube and intake duct clamps and remove the intake duct retainer **(see illustration)**

9.4 Remove the trim retainer from the duct (left arrow) and loosen the duct and carburetor clamps (right arrows)

9.5 Unbolt the bracket from the underside (arrow)
and lift the housing out

9.8 Disconnect the 250 breather hose from the engine

5   Detach the bracket near the driveshaft boot and remove the housing **(see illustration)**.
6   Installation is the reverse of the removal steps.

250 models

Refer to illustrations 9.8 and 9.9

7   Remove the seat and side covers (see Chapter 7).
8   Disconnect the breather hose from the left rear corner of the
engine **(see illustration)**.
9   Loosen the duct clamps at the carburetor and air cleaner housing
(see illustration). Unhook the retaining clips and lift the air cleaner
housing out of the vehicle.
10   Installation is the reverse of the removal steps.

10   Throttle cable and housing - removal, installation and adjustment

1   The throttle on these vehicles is operated by a thumb lever on the
right handlebar.

Removal

350 models

Refer to illustrations 10.3a and 10.3b

2   Remove the carburetor (see Section 6).
3   Remove the throttle housing cover from the carburetor **(see illustration)**. Loosen the locknut and adjusting nut to create slack in the
cable, then rotate the throttle pulley, align the cable with the pulley slot
and slip the cable out of the pulley **(see illustration)**. Slip the cable out
of the bracket slot to complete disconnection at the carburetor end.

9.9 Loosen the clamps at the carburetor and air cleaner housing
(250 models)

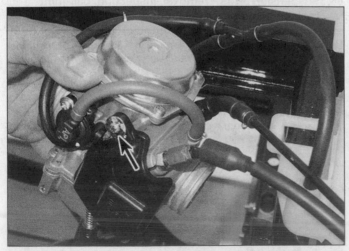

10.3a Remove the throttle pulley cover screw (arrow)
and lift off the cover

10.3b Loosen the locknut and adjusting nut (arrows) and slip the
cable out of the slot, then align the cable with the slot in
the throttle pulley and slip it out

10.5a Remove the screws (arrows) and lift the cover and gasket off the throttle housing

10.5b Bend back the lockwasher tab (arrow), remove the nut and lockwasher . . .

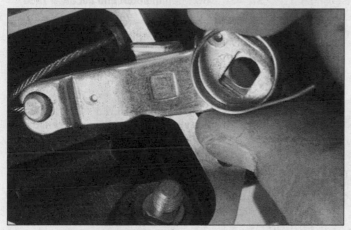

10.5c . . . then lift off the lever and spring . . .

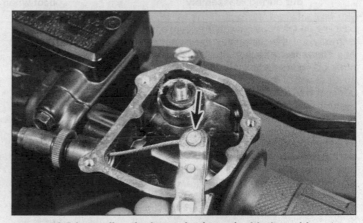

10.5d . . . align the lever slot (arrow) with the cable and slip the cable out

250 models

4   Unscrew the carburetor top and separate the jet needle and throttle cable from the throttle piston (see Section 7).

All models

Refer to illustrations 10.5a, 10.5b, 10.5c, 10.5d and 10.7

5   Remove the cover from the throttle housing on the handlebar, remove the lever components and disconnect the cable **(see illustrations)**.

6   Remove the cable, noting how it's routed.

7   If necessary, remove the throttle housing clamp screws and detach the throttle housing from the handlebar **(see illustration)**.

Installation

Refer to illustration 10.8

8   If the throttle housing was removed, install it on the handlebar and tighten its clamp screws loosely. Position the housing so its outer end is aligned with the handlebar punch mark **(see illustration)**. Align the

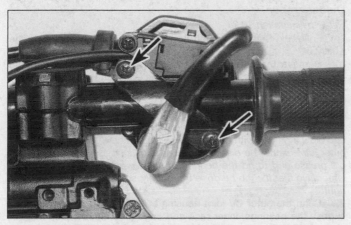

10.7 Remove the clamp screws (arrows) to detach the throttle housing

10.8 Align the punch mark on the handlebar (under the throttle housing, right arrow) with the outer end of the throttle housing; align the throttle housing line with the brake master cylinder seam (left arrows)

11.3a Loosen the clamp and disconnect the cable from the bracket . . .

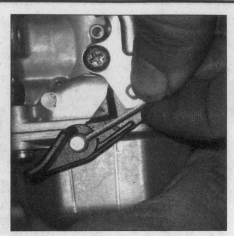

11.3b . . . rotate the cable to align it with the slot and slip it out

11.4a Unscrew the nut to detach the knob/cable assembly from the handlebar bracket . . .

cast line on the housing with the seam of the brake master cylinder and clamp. Install the clamp and screws and tighten them securely.

9   Route the cable into place. Make sure it doesn't interfere with any other components and isn't kinked or bent sharply.

10   Lubricate the end of the cable with multi-purpose grease and connect it to the throttle pulley at the carburetor. Pass the inner cable through the slot in the bracket, then seat the cable housing in the bracket. Install the carburetor on the engine (see Section 6). Make sure all cable retainers are securely installed.

11   Reverse the disconnection steps to connect the throttle cable to the handlebar lever, noting how the spring is installed **(see illustration 10.5c)**. Operate the lever and make sure it returns to the idle position by itself under spring pressure. **Warning:** *If the lever doesn't return by itself, find and solve the problem before continuing with installation. A stuck lever can lead to loss of control of the vehicle.*

Adjustment

12   Follow the procedure outlined in Chapter 1, Throttle operation/ grip freeplay - check and adjustment, to adjust the cable.

13   Turn the handlebars back and forth to make sure the cables don't cause the steering to bind.

14   Once you're sure the cable operates properly, install the covers on the throttle housing and carburetor.

15   With the engine idling, turn the handlebars through their full travel (full left lock to full right lock) and note whether idle speed increases. If it does, the cable is routed incorrectly. Correct this dangerous condition before riding the vehicle.

11 Choke cable - removal and installation

Removal

Refer to illustrations 11.3a, 11.3b, 11.4a and 11.4b

1   Remove the seat and fuel tank (see Chapter 7 and Section 2).

2   If you're working on a 350 model, unscrew the cable, together with the starting enrichment valve, from the carburetor **(see illustration 6.5)**.

3   If you're working on a 250 model, loosen the screw and free the choke cable from the lever **(see illustrations)**.

4   Follow the choke cable to the knob, noting how it's routed, and free it from its retainers. Unscrew the nut to detach the knob from the bracket **(see illustrations)** and remove the knob and cable as an assembly.

Installation

5   Installation is the reverse of the removal steps. Make sure the cable is securely fastened in its clips.

6   Install the fuel tank and all of the other components that were previously removed.

12 Exhaust system - removal and installation

Refer to illustrations 12.2, 12.4a, 12.4b, 12.6a and 12.6b

1   Refer to Chapter 7 and remove the right side cover.

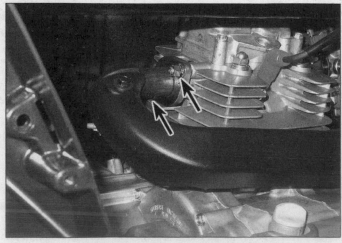

11.4b . . . or from the bracket near the instrument panel

12.2 Remove the holder nuts (arrows) and slip the holder off the studs

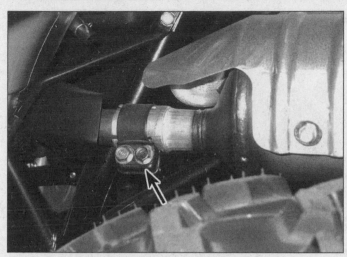

12.4a Loosen the muffler clamp bolts (arrow) . . .

12.4b . . . and remove the muffler mounting bolts (arrows)

2   Remove the exhaust pipe holder nuts and slide the holder off the mounting studs **(see illustration)**.
3   If necessary, unbolt the heat shield and remove it from the exhaust pipe.

4   Loosen the muffler clamp bolts and remove the mounting bolts **(see illustrations)**.
5   Pull the exhaust system forward, separate the pipe from the cylinder head and remove the system from the machine.
6   Installation is the reverse of removal, with the following additions:

  a) *Be sure to install new gaskets at the muffler clamp and cylinder head* **(see illustrations)**.
  b) *Tighten the muffler mounting bolts and heat shield bolts securely, but don't overtighten them and strip the threads.*

12.6a Install a new gasket at the muffler joint . . .

13 Carburetor heater (350 models) - removal and installation

Refer to illustration 13.3
1   Disconnect the carburetor heater connector **(see illustration 6.4)**.
2   Connect an ohmmeter between the terminals and measure the resistance. If it's not within the range listed in this Chapter's Specifications, replace the carburetor heater as described below.
3   With rags handy to catch spilled fuel, unscrew the carburetor heater from the float bowl **(see illustration)**. Remove the heater and spacer.
4   Installation is the reverse of the removal steps. Use a new sealing washer.

12.6b . . . and in the exhaust port

13.3 Unscrew the carburetor heater and remove the collar (arrow)

Notes

Chapter 4
Ignition system

Contents

	Section
Engine kill switch - check, removal and installation	See Chapter 8
General information	1
Ignition (main) switch and key lock cylinder - check, removal and installation	See Chapter 8
Ignition coil - check, removal and installation	3

	Section
Ignition control module (ICM) - harness check, removal and installation	5
Ignition system - check	2
Ignition timing - general information and check	6
Pulse generator - check, removal and installation	4
Spark plug replacement	See Chapter 1

Specifications

Ignition coil peak voltage (minimum)	100 volts
Pulse generator peak voltage (minimum)	0.7 volts
Ignition timing	
At idle	13-degrees BTDC
At full advance (4500 rpm)	30-degrees BTDC

Torque specifications

Pulse generator screws or Allen bolts	6 Nm (52 inch-lbs)*
Timing hole cap	See Chapter 1

*Apply non-permanent thread locking agent to the bolt threads.

1   General information

These vehicles are equipped with a battery operated, fully transistorized, breakerless ignition system. The system consists of the following components:

Pulse generator
Ignition control module (ICM)
Battery and fuse
Ignition coil
Spark plug
Engine kill (stop) and main (key) switches
Primary and secondary (HT) circuit wiring

The transistorized ignition system functions on the same principle as a DC ignition system with the pulse generator and ICM performing the tasks previously associated with the breaker points and mechanical advance system. As a result, adjustment and maintenance of ignition components is eliminated (with the exception of spark plug replacement).

Because of their nature, the individual ignition system components can be checked but not repaired. If ignition system troubles occur, and the faulty component can be isolated, the only cure for the problem is to replace the part with a new one. Keep in mind that most electrical parts, once purchased, can't be returned. To avoid unnecessary expense, make very sure the faulty component has been positively identified before buying a replacement part.

2   Ignition system - check

Refer to illustrations 2.5 and 2.13
Warning: *Because of the very high voltage generated by the ignition system, extreme care should be taken when these checks are performed.*

1   If the ignition system is the suspected cause of poor engine performance or failure to start, a number of checks can be made to isolate the problem.

2   Make sure the ignition kill (stop) switch is in the Run or On position.

Engine will not start

3   Refer to Chapter 1 and disconnect the spark plug wire. Connect the wire to a spare spark plug and lay the plug on the engine with the threads contacting the engine. If necessary, hold the spark plug with an insulated tool. Crank the engine over and make sure a well-defined, blue spark occurs between the spark plug electrodes. **Warning:** *Don't remove the spark plug from the engine to perform this check - atomized fuel being pumped out of the open spark plug hole could ignite, causing severe injury!*

4   If no spark occurs, the following checks should be made:

5   Unscrew the spark plug cap from the plug wire and check the cap resistance with an ohmmeter **(see illustration)**. If the resistance is infinite, replace it with a new one.

6   Make sure all electrical connectors are clean and tight. Check all wires for shorts, opens and correct installation.

7   Check the battery voltage with a voltmeter. If the voltage is less than 12-volts, recharge the battery.

8   Check the ignition fuse and the fuse connections (see Chapter 8). If the fuse is blown, replace it with a new one; if the connections are loose or corroded, clean or repair them.

9   Refer to Section 3 and check the ignition coil primary and secondary resistance.

10   Refer to Section 4 and check the pulse generator resistance.

11   If the preceding checks produce positive results but there is still no spark at the plug, refer to Section 5 and check the ICM.

Engine starts but misfires

12   If the engine starts but misfires, make the following checks before deciding that the ignition system is at fault.

13   The ignition system must be able to produce a spark across a seven millimeter (1/4-inch) gap (minimum). A simple test fixture **(see illustration)** can be constructed to make sure the minimum spark gap can be jumped. Make sure the fixture electrodes are positioned seven millimeters apart.

14   Connect one of the spark plug wires to the protruding test fixture electrode, then attach the fixture's alligator clip to a good engine ground (earth).

15   Crank the engine over with the key in the On position and see if well-defined, blue sparks occur between the test fixture electrodes. If the minimum spark gap test is positive, the ignition coil is functioning properly. If the spark will not jump the gap, or if it is weak (orange colored), refer to Steps 5 through 11 of this Section and perform the component checks described.

---

3   Ignition coil - check, removal and installation

Check

1   In order to determine conclusively that the ignition coils are defective, they should be tested by an authorized Honda dealer service department which is equipped with the special electrical tester required for this check.

2   Honda doesn't provide ignition coil resistance specifications for these models. Instead, it recommends testing the coil peak primary voltage with one of the following:

a) *Imrie diagnostic tester model 625*
b) *Digital multimeter (minimum impedance 10 meg-ohms per DC volt) and peak voltage adapter (Honda part no. 07HGJ-0020100 or equivalent)*

If you don't have one of these special tools, have the coil tested by a Honda dealer. If you do have the right equipment, the procedure is as follows.

3   Place the transmission in Neutral and disconnect the spark plug wire from the plug (see Chapter 1). Connect another spark plug to the cap and lay it on a bare metal part of the engine so it's grounded.

4   Leave the ignition coil wires connected for this step. Connect the peak voltage tester negative terminal to the ignition coil black/yellow wire terminal. Connect the tester positive terminal to body ground (bare

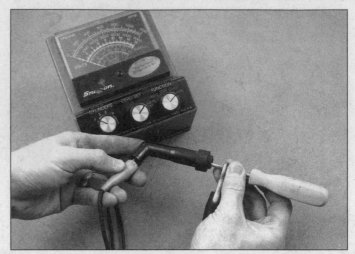

2.5 Unscrew the spark plug cap from the plug wire and measure its resistance with an ohmmeter

metal on the engine). The meter must be on its "10 M-ohms/DCV" setting or it won't produce an accurate or meaningful measurement.

5   Set the engine start switch to Run and crank the engine with the starter. Compare the tester reading to the value listed in this Chapter's Specifications.

6   There are four possible outcomes: peak voltage is adequate, peak voltage is too low, there is no voltage, or peak voltage is satisfactory but there is still no spark at the spark plug.

a) *If peak voltage is adequate, the ignition coil is operating satisfactorily.*
b) *If peak voltage is low, go to Step 7.*
c) *If there is no voltage , go to Step 15.*
d) *If peak voltage is satisfactory, but there is no spark at the plug, go to Step 20.*

Peak voltage is low

7   First, check the peak voltage adapter connections by repeating Step 5 with the tester connections reversed. If the indicated voltage is now above the specified minimum, the coil is OK (the adapter connections were incorrect the first time you measured coil peak voltage).

8   Check your multimeter's impedance setting. It should be at "10 M-ohms/DCV." If it isn't, change to the correct setting and repeat Step 5.

9   If the battery is undercharged, the cranking speed might be too

2.13 A simple spark gap testing fixture can be made from a block of wood, two nails, a large alligator clip, a screw and a piece of wire

3.24 Locate the coil on the left side of the vehicle . . .

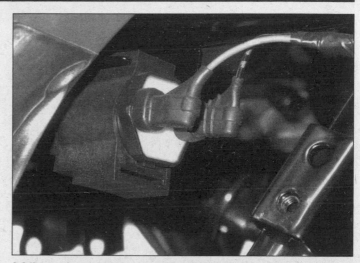

3.24b . . . then label and disconnect the primary wires and remove the coil (350 shown; 250 similar)

low. Charge the battery (see Chapter 8).

10   The sampling timing of the tester and measured pulse are not synchronized. The ignition system is operating satisfactorily if the measured peak voltage is higher than the specified minimum peak voltage at least once (this only applies to the peak voltage tester, not to a multimeter used with the peak voltage adapter).

11   There is a bad connection or broken wire in the ignition system (see the wiring diagrams at the end of Chapter 8). Check the wiring between the alternator and ICM.

12   The ignition pulse generator is defective (see Section 4).

13   The ignition coil is defective (substitute a known good coil and recheck).

14   If everything in Steps 7 through 13 is satisfactory, the ignition control module is defective (see Section 6).

No peak voltage

15   Verify the tester connections and setting as described in Steps 7 and 8 above.

16   Test the engine kill switch and ignition switch (see Chapter 8).

17   Check the ICM green/white (ground) wire for a break or poor connection.

18   Check the ignition pulse generator (see Section 4).

19   If everything in Steps 15 through 18 is OK, the ignition coil module may be defective (see Section 5).

Peak voltage satisfactory but no spark at the plug

20   The spark plug or plug wire may be defective (see Chapter 1).

21   Secondary ignition current might be "leaking" somewhere between the coil and spark plug.

22   If the plug and wire are good and there's no secondary current leak, the ignition coil is probably defective. Substitute a known good coil or have it tested by a Honda dealer before replacing it.

Removal and installation

Refer to illustrations 3.24a and 3.24b

23   Refer to Chapter 3 and remove the fuel tank and heat shield.

24   Disconnect the spark plug wire from the plug. After labeling them with tape to aid in reinstallation, disconnect the coil primary circuit electrical connectors **(see illustrations)**.

25   Slip the rubber coil mount off its blade, then remove the coil from the mount.

26   Installation is the reverse of removal. Make sure the primary circuit electrical connectors are attached to the proper terminals. Just in case you forgot to mark the wires, the black/yellow wire connects to the positive terminal and the green wire connects to the negative terminal.

4   Pulse generator - check, removal and installation

Check

Refer to illustration 4.2

1   Testing the pulse generator on these models is done by checking the peak voltage. It requires the special tester described in Section 3.

2   Locate the four-pin connector at the ignition control module at the front of the vehicle **(see illustration)**. Disconnect the connector.

3   Connect the tester positive probe to the connector blue-yellow wire terminal and the negative probe to the green/white wire terminal. **Note:** *Leave the wiring connector connected. Connect the tester by backprobing the terminals on the harness side of the connector.*

4   Place the transmission in Neutral and disconnect the spark plug wire from the plug (see Chapter 1). Connect another spark plug to the cap and lay it on a bare metal part of the engine so it's grounded.

5   Set the engine start switch to Run and crank the engine with the starter. Compare the peak voltage reading to the value listed in this Chapter's Specifications.

6   If voltage is less then the peak voltage listed in this Chapter's Specifications, locate the 5-pin round alternator connector on the right side of the vehicle beneath the rear fender. Disconnect the connector, then connect the tester to the blue/yellow and green/white wire terminals in the alternator side of the connector.

7   Measure peak voltage as described in Step 4.

4.2 The ignition control module is mounted under the front fender at the center of the vehicle (arrow) (350 shown; 250 similar)

8   If the reading in Step 4 (at the ICM connector) is not within the Specifications, but the reading in Step 6 (at the alternator connector) is within the Specifications, check the wiring for a break or poor connection, referring to the wiring diagrams in Chapter 8,.

9   If the reading is incorrect at both connectors (ICM and alternator), perform Steps 7 through 13 in Section 3. If these steps don't locate the problem, replace the pulse generator.

Removal

10   Remove the alternator cover and disconnect the pulse generator connector, if you haven't already done so (see Chapter 8).

11   Free the pulse generator wire from the groove in the pulse generator body.

12   Unscrew the pulse generator Allen bolts and remove the pulse generator.

Installation

13   Installation is the reverse of the removal procedure, with the following addition: Apply non-permanent thread locking agent to the pulse generator screws or Allen bolts and tighten them to the torque listed in this Chapter's Specifications.

14   Refer to Chapter 8 and reinstall the alternator cover.

5   Ignition control module (ICM) - harness check, removal and installation

1   The ICM can only be checked by process of elimination; that is, after all other possible causes of ignition problems have been checked and eliminated, the ICM is probably at fault.

2   Perform the checks in Sections 2 through 4 to test other parts of the ignition system. If all of these test OK, the ICM is probably at fault. However, since a new ICM can't be returned once purchased, it's a good idea to have a dealer service department or other qualified shop check the ignition system before buying a new ICM.

6   Ignition timing - general information and check

General information

1   Ignition timing need be checked only if you're troubleshooting a problem such as loss of power. Since the ignition timing can't be adjusted and since none of the ignition system parts is subject to mechanical wear, there's no need for regular checks.

2   The ignition timing is checked with the engine running, both at idle and at a higher speed listed in this Chapter's Specifications. Inexpensive neon timing lights should be adequate in theory, but in practice may produce such dim pulses that the timing marks are hard to see. If possible, one of the more precise xenon timing lights should be used, powered by an external source of the appropriate voltage. **Note:** *Don't use the vehicle's own battery as an incorrect reading may result from stray impulses within the electrical system.*

Check

3   Warm the engine to normal operating temperature, make sure the transmission is in Neutral, then shut the engine off.

4   Refer to Valve clearance - check and adjustment in Chapter 1 and remove the timing window cap.

5   Connect the timing light and a tune-up tachometer to the engine, following manufacturer's instructions.

6   Start the engine. Make sure it idles at the speed listed in the Chapter 1 Specifications. Adjust if necessary.

7   Point the timing light into the timing window. At idle, the line next to the F mark on the alternator rotor should align with the notch at the side of the timing window.

8   Raise engine by turning the throttle stop screw (see Chapter 1). The timing mark on the alternator rotor should move in relation to the notch.

9   If the timing is incorrect and all other ignition components have tested as good, the ICM may be defective. Have it tested by a dealer service department or other qualified shop.

10   When the check is complete, grease the timing window O-ring, then install the O-ring and cap and disconnect the test equipment.

Chapter 5
Steering, suspension and final drive

Contents

	Section			Section
General information	1	Front driveaxle (4WD models) - boot replacement		
Handlebars - removal, inspection and installation	2	and CV joint overhaul		10
Steering shaft - removal, inspection, bearing replacement		Front differential and driveshaft - removal, inspection		
and installation	3	and installation		11
Shock absorbers - removal and installation	4	Rear axle shaft and housing - removal, inspection and installation		12
Tie-rods - removal, inspection and installation	5	Rear differential - removal, inspection and installation		13
Steering knuckles - removal, inspection, and installation	6	Swingarm bearings - check		14
Steering knuckle bearing replacement	7	Swingarm - removal and installation		15
Suspension arms - removal, inspection, ball-joint		Swingarm bearings - replacement		16
replacement and installation	8	Rear driveshaft - removal, inspection and installation		17
Front driveaxles (4WD models) - removal and installation	9			

Specifications

Tie-rod ball-joint spacing
Rancher 2WD models .... 355+/-1 mm (14+/-0.04 inches)
Rancher 4WD models .... 346+/-1 mm (13.6+/-0.04 inches)
Recon models .... 323+/-1mm (12.7+/-0.04 inches)
TRX250EX models .... 326.7+/-1 mm (12.86+/-0.04 inches)
Tie-rod exposed thread length
Rancher 2WD models
2000 through 2003 .... 4.5 mm (0.18 inch)
2004 and later... 5 mm (0.20 inch)
Rancher 4WD models .... 8.75 mm (0.3 inch)
Recon and TRX250EX models... 5.5 mm (0.22 inch)
Rear axle runout (maximum) .... 3.0 mm (0.12 inch)
Rear axle bearing depth
TRX350 models.. 11 mm (0.43 inch)
TRX250 models.. 12 to 13 mm (0.47 to 0.51 inch)

Torque specifications

Grip end bolts... Not specified
Handlebar bracket bolts .... Not specified
Handlebar bracket lower nuts .... 39 Nm (29 ft-lbs) (1)
Tie-rod stud nuts .... 54 Nm (40 ft-lbs) (1)
Tie-rod locknuts .... 54 Nm (40 ft-lbs)
Steering shaft nut .... 108 Nm 80 ft-lbs)
Steering shaft upper bracket bolts.. 32 Nm (24 ft-lbs)
Front shock absorbers (1)
Upper nut and bolt... 30 Nm (22 ft lbs)
Lower nut and bolt
Rancher and TRX250EX models .... 30 Nm (22 ft-lbs)
Recon models
1997 .... 54 Nm (40 ft-lbs)
1998 and later .... 49 Nm (36 ft-lbs)

Torque specifications

Rear shock absorbers
    Upper bolt/nut.. 44 Nm (33 ft-lbs) (1)
    Lower bolt
        Rancher and TRX250EX models .... 44 Nm (33 ft-lbs)
        Recon models.. 54 Nm (40 ft-lbs)
Front suspension arm pivot bolt nuts (1)
    Rancher models... 44 Nm (33 ft-lbs)
    Recon and TRX250EX models .... 30 Nm (22 ft-lbs)
Ball-joint nuts
    Rancher and Recon models .... 30 Nm (22 ft-lbs)
    TRX250EX models
        2005 and earlier.. 32 Nm (24 ft-lbs)
        2006 and later... 29 Nm (21 ft-lbs)
Front differential (Rancher 4WD models)
    Front bracket to differential mounting bolt/nut .... 22 Nm (16 ft-lbs)
    Upper mounting bolts/nuts... 44 Nm (33 ft-lbs)
Rear axle and differential (Rancher models)
    Axle housing to differential nuts.................................... 44 Nm (33 ft-lbs)
    Rear differential to swingarm bolts .... 54 Nm (40 ft-lbs)
    Rear differential skid plate bolts.................................... 32 Nm (24 ft-lbs)
Rear axle and differential (Recon and TRX250EX models)
    Axle shaft nut... 39 Nm (29 ft-lbs)
    Axle shaft locknut... 127 Nm (94 ft-lbs)
    Differential to swingarm bolts .... 54 Nm (40 ft-lbs)
    Differential skid plate bolts.. Not specified
Swingarm
    Rancher models
        Left pivot bolt... 118 Nm (87 ft-lbs)
        Right pivot bolt .... 4 Nm (36 inch-lbs)
        Right pivot bolt locknut .... 118 Nm (87 ft-lbs)
    Recon and TRX250EX models
        Right pivot bolt .... 113 Nm (83 ft-lbs)
        Left pivot bolt... 4 Nm (36 inch-lbs)
        Left pivot bolt locknut... 113 Nm (83 ft-lbs)

1   *Use new nuts; don't reuse the old ones.*

1   General information

The front suspension on all models consists of an upper and lower control arm on each side of the vehicle. A shock absorber with concentric coil spring is installed between the upper suspension arm and the frame.

The rear suspension on all models consists of a shock absorber with concentric coil spring and a steel swingarm. Final drive is by a shaft, which passes through an integral tube on the swingarm.

The steering system consists of knuckles mounted at the outer ends of the front suspension and connected to a steering shaft by tie rods. The steering shaft is turned by handlebars.

2   Handlebars - removal, inspection and installation

1   The handlebars are a one-piece tube. The tube fits into a bracket, which is integral with the steering shaft. If the handlebars must be removed for access to other components, such as the steering shaft, simply remove the bolts and slip the handlebars off the bracket. It's not necessary to disconnect the cables, wires or brake hose, but it is a good idea to support the assembly with a piece of wire or rope, to avoid unnecessary strain on the cables, wires and the brake hose.

2   If the handlebars are to be removed completely, refer to Chap-
ter 3 for the throttle housing removal procedure, Chapter 6 for the master cylinder removal procedure and Chapter 8 for the switch removal procedure.

Removal

Refer to illustrations 2.3, 2.5 and 2.6

3   If you plan to remove one of the grips, remove its bolt (all except 2008 and later Recon) or end cap (2008 and later Recon). To remove an end cap, twist it with an Allen wrench to free its retaining tabs from their holes inside the handlebar, then pull or pry the end cap out of the handlebar.

4   If you're working on a US 350 model or any 250 model, remove the handlebar cover. If you're working on a non-US 250 model, remove the instrument cluster (see Chapter 8). On all models, detach the fuel tank breather hose.

5   Remove the handlebar bracket bolts and lift off the brackets **(see illustration)**. Lift the handlebar out of the bracket.

6   To remove the brackets, remove their mounting nuts and washers from below **(see illustration)**. Discard the nuts and use new ones on installation.

Inspection

7   Check the handlebar and brackets for cracks and distortion and replace them if any undesirable conditions are found.

2.3 The grips are secured to the handlebars by a bolt (shown) or an end cap and adhesive

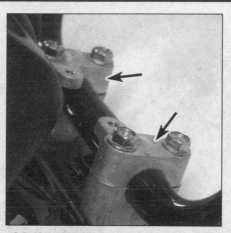

2.5 Unbolt the brackets (arrows) to free the handlebar

2.6 Remove the bracket nuts and washers (arrows); discard the nuts and use new ones on installation

2.8 Line up punch mark on the handlebar (arrow) with the bracket parting line

3.3a Remove the bolts (arrows) and lift off the bracket . . .

3.3b . . . to expose the bushing; the Up mark and arrow must be upright when the bushing is installed

Installation

Refer to illustration 2.8

8   Installation is the reverse of the removal steps, with the following additions:

a) *If the bracket nuts were removed, install new nuts and tighten them to the torque listed in this Chapter's Specifications.*

b) *When installing the handlebar to the brackets, line up the punch mark on the handlebar with the bracket seams (see illustration). Tighten the front bracket bolts securely, then tighten the rear bracket bolts. There will be a small gap at the rear of each bracket (between the upper and lower parts of the bracket) when the bolts are correctly tightened. Don't try to close the gap by over-tightening the rear bolts, or the brackets will break.*

c) *If a grip was removed, clean the handlebar and apply cement (Honda Hand Grip Cement or equivalent) to the inside of the grip. Let the cement cure for three to five minutes, then slide the grip onto the handlebar and twist it to spread the cement. Let the cement dry for an hour, then install the grip end bolt.*

3   Steering shaft - removal, inspection, bearing replacement and installation

Removal

Refer to illustrations 3.3a, 3.3b, 3.5a and 3.5b

1   Remove the handlebars and their bracket (Section 2).

2   Remove the front fender (see Chapter 7).

3   Unbolt the steering shaft bracket from the frame to expose the bushing **(see illustrations)**.

4   Refer to Section 5 and disconnect the inner ends of the tie-rods.

5   Remove the cotter pin and steering shaft nut **(see illustration)**. On TRX350 models, remove the steering arm from the shaft **(see illustration)**.

6   Remove the steering shaft from the vehicle.

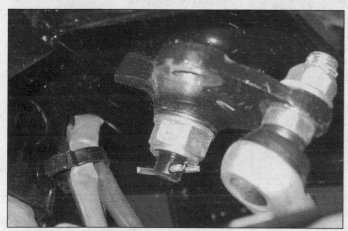

3.5a Remove the cotter pin and nut, then detach the steering arm from the shaft

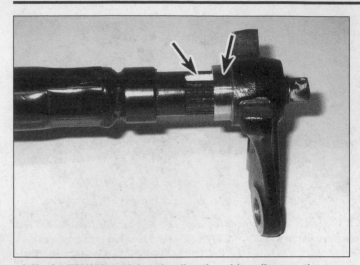

3.5b On TRX350 models only, align the wide splines on the arm and shaft (arrows) when installing the arm

3.11a Pry out the seal above the bearing . . .

Inspection

7   Clean all the parts with solvent and dry them thoroughly, using compressed air, if available.

8   Check the steering shaft bushing for wear, deterioration or damage **(see illustration 3.3b)**. Replace it if there's any doubt about its condition.

9   Check the steering shaft for bending or other signs of damage. Inspect the steering arm as well. Do not attempt to repair any steering components. Replace them with new parts if defects are found.

10   Insert a finger into the steering shaft bearing and turn the inner race. If it's rough, loose or noisy, replace it as described below.

Bearing replacement

Refer to illustrations 3.11a, 3.11b, 3.12 and 3.14

11   Pry the upper and lower grease seals out of the bearing housing in the frame **(see illustrations)**.

12   Remove the snap-ring from on top of the bearing **(see illustration)**, then remove the bearing through the top of the housing.

13   Pack a new bearing with high-quality grease (preferably a moly-based grease). Position the bearing in the housing with its sealed side up and tap it into place with a bearing driver or socket that bears against the bearing outer race.

14   Install the snap-ring. Coat the lips of new upper and lower grease seals with grease, then install them in the housing **(see illustration)**.

Installation

15   Installation is the reverse of removal, with the following additions:

a) *Pack the steering shaft bushing's internal cavities with grease. Install it with its UP mark upward* **(see illustration 3.3b)**.

b) *On TRX350 models, align the wide splines on the steering arm and shaft, then install the steering arm (see illustration 3.5b).*

c) *Lubricate the threads and flange of the steering shaft nut with grease.*

d) *Use new cotter pins and tighten all fasteners to the torques listed in this Chapter's Specifications.*

---

4   Shock absorbers - removal and installation

Front shock absorbers

Refer to illustration 4.2

1   Securely block both rear wheels so the vehicle won't roll. Refer to Chapter 6 and remove the front wheels.

2   Remove the mounting nut and bolt from the bottom of the shock, then from the top **(see illustration)**. Lift the shock out. Discard the mounting nuts and use new ones on installation.

3   Installation is the reverse of the removal steps. Use new nuts (don't re-use the old ones) and tighten the nuts and bolts to the torque listed in this Chapter's Specifications.

3.11b . . . and the below the bearing . . .

3.12 . . . and remove the snap-ring, then the bearing

3.14 Press the seals in; finger pressure should be enough, but if not, tap the seals in with a socket the same diameter as the seal

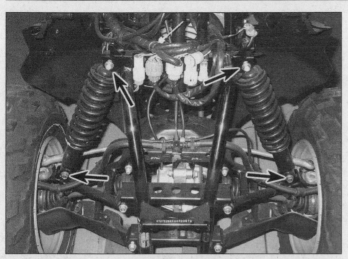

4.2 Remove the nut and bolt from the bottom of the shock absorber, then from the top (arrows); use new nuts on installation and position the bolt heads facing forward

Rear shock absorber

Refer to illustration 4.5

4   Securely block both front wheels so the vehicle can't roll. Jack up the rear end and support it securely with the rear wheels off the ground.

5   Remove the mounting bolts and nuts at the bottom of the shock, then at the top **(see illustration)**. Lift the shock out of the vehicle. Discard the nuts and use new ones on installation.

Rear shock absorber installation

6   Installation is the reverse of the removal steps. Use new nuts. Tighten the nuts and bolts to the torque listed in this Chapter's Specifications.

---

5   Tie-rods - removal, inspection and installation

Removal

Refer to illustrations 5.1, 5.2 and 5.3

1   Remove the cotter pin from the nut at the outer end of the tie-rod **(see illustration)**.

5.2 . . . prevent the stud from turning with an open-end wrench and unscrew the nut, then detach the stud from the steering knuckle

4.5 Remove the nut and bolt from the bottom of the shock absorber, then from the top (arrows)

5.1 Straighten the cotter pin and pull it out of the tie-rod nut . . .

2   Hold the tie-rod flat with a wrench and undo the nut **(see illustration)**. Separate the tie-rod stud from the knuckle.

3   Repeat Steps 1 and 2 to disconnect the inner end of the tie rod **(see illustration)**. Discard the nuts and use new ones on installation.

5.3 Remove the cotter pins and nuts and detach the inner ends of the tie-rods from the steering arm (350 shown; the tie-rod studs on 250 models face downward)

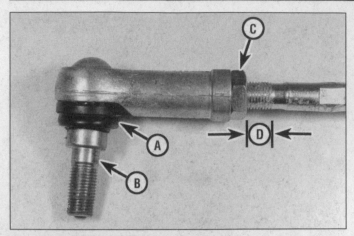

5.5 Check the boot for cracks or deterioration and make sure the stud moves easily, without roughness or looseness

A   Boot
B   Stud
C   Locknut
D   Distance from locknut to end of threads

Inspection

Refer to illustration 5.5

4   Check the tie-rod shaft for bending or other damage and replace it if any problems are found. Don't try to straighten the shaft.

5   Check the ball-joint boot for cracks or deterioration **(see illustration)**. Twist and rotate the threaded stud. It should move easily, without roughness or looseness. If the boot or stud show any problems, unscrew the ball-joint from the tie-rod and install a new one. **Note:** *On Rancher models, the gold-colored ball-joint and locknut go on the end of the tie-rod nearest the wrench flats. On Recon models, the unlabeled ball-joint and gold-colored locknut go on the end of the tie-rod nearest the wrench flats. No specification is available for TRX250EX models.*

Installation

Refer to illustration 5.7

6   Measure the distance from each locknut to the end of the threads and compare it with the value listed in this Chapter's Specifications. If it's incorrect, loosen the locknut and reposition the locknut and ball-joint on the threads.

7   On TRX350 models, position the ball-joint at each end of the tie-rod so the studs face 180-degrees away from each other **(see illustration)**. On TRX250 models, both studs face the same way.

8   The remainder of installation is the reverse of the removal steps, with the following additions:

a) *Use new ball-joint nuts (don't re-use the old ones). Tighten the new nuts to the torque listed in this Chapter's Specifications.*

b) *Use new cotter pins and bend them to hold the nuts securely.*

c) *Check front wheel toe-in and adjust as necessary (see Chapter 1).*

6   Steering knuckles - removal, inspection, and installation

Removal

Refer to illustrations 6.5, 6.6 and 6.7

1   Separating the ball-joint(s) from the knuckle requires either a special Honda separator tool or its equivalent. Equivalent automotive tools can be rented, but they must be small enough for use on these vehicles. An automotive tie-rod separator tool, rather than a ball-joint separator, may be the correct size. If the correct special tools aren't available, the knuckle can be removed as an assembly with the suspension arms. This assembly can then be taken to a Honda dealer for ball-joint removal (and knuckle bearing replacement if necessary).

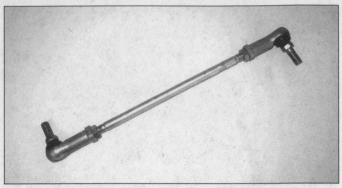

5.7 Position the tie-rod studs at 180-degrees from each other

6.5 Remove the cotter pins and nuts (arrows) . . .

2   Securely block both rear wheels so the vehicle won't roll. Loosen the front wheel nuts with the tires still on the ground, then jack up the front end, support it securely on jackstands and remove the front wheels.

3   Refer to Section 5 and disconnect the outer end of the tie-rod from the knuckle.

4   Remove the front brake panel (drum brake models) or caliper and disc (disc brake models) (see Chapter 6). The front brake hose can be left connected, but be careful not to twist it and be sure to support the panel or caliper with wire or rope so it doesn't hang by the brake hose.

5   Remove the cotter pin and nut from the upper and lower ball-joints **(see illustration)**.

6.6 . . . and detach the ball-joint from the knuckle with the special tool or an aftermarket equivalent (shown)

6.7 Tapping on the boss where the ball-joint stud passes through it is another way to free the ball-joint

6.8 Inspect the ball-joint boot and knuckle bearing (arrows)

6   Separate the lower ball-joint from the knuckle with Honda tool 07MAC-SL00200 or equivalent **(see illustration)**. If you're using the Honda tool, lubricate the puller jaw and the pressure bolt threads with multi-purpose grease. Slip the tool into position, taking care not to damage the rubber boot, and turn the adjusting bolt so the jaws are parallel. Tighten the pressure bolt by hand, make sure the jaws are still parallel (readjust the pressure bolt if necessary), then tighten the pressure bolt with a wrench until the ball-joint stud pops out of the knuckle.
7   If there isn't room to fit the separator tool onto the upper ball-joint, loosen the nut and tap on the knuckle boss with a mallet where the stud passes through it **(see illustration)**. This should loosen the stud so it can be removed from the knuckle.

Inspection
Refer to illustration 6.8
7   Check the knuckle carefully for cracks, bending or other damage. Replace it if any problems are found. If the vehicle has been in a collision or has been bottomed hard, it's a good idea to have the knuckle magnafluxed by a machine shop to check for hidden cracks.
8   Check the ball-joint boot for cracks or deterioration **(see illustration)**. Twist and rotate the threaded stud. It should move easily, without roughness or looseness. If the boot or stud show any problems, refer to Section 8 and replace the ball-joint.
9   If you're working on a 4WD model, turn the knuckle bearing inner race with a finger **(see illustration 6.8)**. If it's rough, loose or noisy,

refer to Section 7 and replace it. If you're working on a 2WD model, refer to Chapter 6 to inspect the wheel bearings. On all models, replace the bearing grease seals if they show signs of leakage or wear.

Installation
10   Installation is the reverse of the removal steps, with the following additions: Use new cotter pins and tighten the ball-joint nut(s) to the torque listed in this Chapter's Specifications.

7   Steering knuckle bearing replacement

1   Pry out the bearing grease seals **(see illustration 6.8)**.
2   Remove the bearing snap-ring from the outer side of the knuckle. Drive the bearing out of the knuckle with a bearing driver or socket that bears against the bearing outer race.
3   Pack a new bearing with grease, then drive it in with the same tool used for removal. Seat the bearing securely, then install the snap-ring and make sure it fits completely into its groove.
4   Tap in new grease seals with a bearing driver or socket the same diameter as the seals. Make sure the seals seat squarely in their bores, then lubricate the seal lips with grease.

8   Suspension arms - removal, inspection, ball-joint replacement and installation

Note: *This procedure describes removal and installation of the upper and lower suspension arms. If you plan to remove only the upper or lower arm, ignore the steps which don't apply.*

Removal
Refer to illustrations 8.3, 8.6a and 8.6b
1   Securely block both rear wheels so the vehicle won't roll. Loosen the front wheel nuts with the tires still on the ground, then jack up the front end, support it securely on jackstands and remove the front wheels.
2   Refer to Section 7 and detach the steering knuckle from the suspension arm(s). The tie-rod doesn't have to be detached from the knuckle.
3   Detach the brake hose retainer and breather hose (if equipped) from the upper suspension arm **(see illustration)**.
4   Note which way the lower shock absorber nut faces and remove it **(see illustration 8.3)**.
5   Remove the bolt and detach the lower end of the shock absorber from the upper arm.

8.3 Detach the brake hose retainer, breather hose retainer and the shock absorber lower end from the suspension arm (arrows)

8.6a Note which way the bolt heads face and remove the upper arm pivot bolts and nuts (arrows) . . .

6   Remove the nuts and bolts at the inner end of the upper and lower suspension arms **(see illustrations)** and pull the suspension arms out. Discard the nuts and use new ones on installation.

Inspection

7   Check the suspension arm(s) for bending, cracks or other damage. Replace damaged parts. Don't attempt to straighten them.
8   Check the rubber bushings at the inner end of the suspension arm for cracks, deterioration or wear of the metal insert. Check the pivot bolts for wear as well. Replace the suspension arm if any problems are visible.
9   Check the ball-joint boot for cracks or deterioration. Twist and rotate the threaded stud. It should move easily, without roughness or looseness.

Ball-joint replacement

Refer to illustrations 8.10a and 8.10b
Note: *This procedure requires a press and special support tools (Honda 07WMF-HN00100 and 07946-1870100 or equivalent). If you don't have them, take the knuckle to a Honda dealer for ball-joint replacement.*
10   Remove the ball-joint snap-ring **(see illustrations)**.
11   If you're working on the lower ball-joint of a 4WD model, place the knuckle in a press with the ball-joint stud upward. Place the driver portion of the removal tool under the ball-joint and the hollow portion over the stud and press the ball-joint out.

12   Place the new ball-joint on the press plate and place the knuckle over the stud.
13   Place the support tool over the stud so the stud will fit into the tool when the ball-joint is pressed on. Press the ball-joint into the knuckle; if it won't go easily, stop and make sure the support tool is aligned correctly.
14   Install the snap-ring and make sure it seats completely in its groove.
15   If you're working on a suspension arm ball-joint (upper or lower), remove the snap-ring **(see illustration 8.10b)**, then replace the lower ball-joint using the methods and tools described above.

Installation

16   Installation is the reverse of the removal steps, with the following additions:

a) *Use new nuts on the suspension arm pivot bolts and the shock absorber lower bolt.*
b) *Tighten the nuts and bolts slightly while the vehicle is jacked up, then tighten them to the torque listed in this Chapter's Specifications while the vehicle's weight is resting on the wheels.*

---

9   Front driveaxles (4WD models) - removal and installation

Removal

Refer to illustrations 9.5, 9.6a and 9.6b
1   Securely block both rear wheels so the vehicle won't roll. Loosen the front wheel nuts with the tires still on the ground, then jack up the front end, support it securely on jackstands and remove the front wheels.
2   Remove the front brake drum and disconnect the breather tube from the brake panel (see Chapter 5).
3   Remove the cotter pin from the lower ball-joint nut, then loosen the nut but don't remove it yet.
4   Separate the lower ball-joint from the knuckle (see Section 6).
5   Freeing the outer end of the driveaxle from the knuckle requires that the knuckle be pulled outward off of the outer end of the driveaxle. If you have the necessary ball-joint separator tool, you can use it to separate the ball-joint studs from the knuckle, which will free the knuckle to be pulled outward (see Section 6). If you don't have the ball-joint tool, the inner ends of the suspension arms and the lower end of the shock absorber can be unbolted (see Sections 4 and 8) and the suspension arms and knuckle pulled outward as a unit **(see illustration)**.
6   The inner end of the driveaxle is held in the final drive unit by a circlip **(see illustration)**. With the outer end of the driveaxle free of the

8.6b . . . and do the same thing to remove the lower arm

8.10a Remove the ball-joint snap-ring (arrow)

8.10b Remove the snap-ring (arrow), then press out the lower ball-joint

9.5 The suspension arms and knuckle can be removed together if you don't have a ball-joint separator tool

knuckle, grasp the joint at the inner end firmly so it won't be pulled apart, then pry the driveaxle out of the final drive unit **(see illustration)**. **Caution:** *Pull the driveaxle straight out (don't let it tilt up, down or sideways) to prevent damage to the oil seal in the final drive unit.*

Installation

7   Installation is the reverse of the removal steps. After installing the driveaxle in the final drive unit, tug outward on it to make sure the circlip is locked in place. Tighten the hub nut to the torque listed in the Chapter 6 Specifications.

10 Front driveaxle (4WD models) - boot replacement and CV joint overhaul

Inner CV joint and boot

Disassembly

Refer to illustrations 10.3, 10.4, 10.6 and 10.8

1   Remove the driveaxle from the vehicle (see Section 9).
2   Mount the driveaxle in a vise. The jaws of the vise should be lined with wood or rags to prevent damage to the axleshaft.
3   Pry the boot clamp retaining tabs up with a small screwdriver and slide the clamps off the boot **(see illustration)**.
4   Slide the boot back on the axleshaft and pry the wire ring ball retainer from the outer race **(see illustration)**.
5   Pull the outer race off the inner bearing assembly.
6   Remove the snap-ring from the groove in the axleshaft with a pair of snap-ring pliers **(see illustration)**.
7   Slide the inner bearing assembly off the axleshaft.

9.6a Pull on the joint (not on the shaft) to free the circlip (arrow) from the differential . . .

9.6b . . . and use a pry bar to help separate the joint from the differential

10.3 Pry the boot clamp retaining tabs (arrow) up with a small screwdriver, open the clamps and slide them off the boot

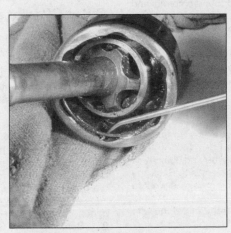

10.4 Pry the wire ring ball retainer out of the outer race

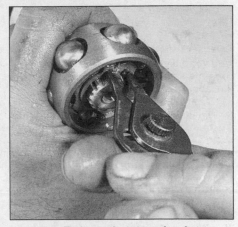

10.6 Remove the snap-ring from the end of the axle

10.8 Apply match marks on the bearing (arrows) to identify which side faces out during reassembly

10.10 Wrap the splined area of the axle with tape to prevent damage to the boot when installing it

10.16 Equalize the pressure inside the boot by inserting a small, dull screwdriver between the boot and the outer race

10.17a To install the new clamps, bend the tang down and . . .

10.17b . . . fold the tabs over to hold it in place

10.21 Slide the boot away from the joint and off the axleshaft

8   Make match marks on the inner and outer portions of the bearing to identify which side faces out on assembly **(see illustration)**.

Inspection

9   Clean the components with solvent to remove all traces of grease. Inspect the cage, balls and races for pitting, score marks, cracks and other signs of wear and damage. Shiny, polished spots are normal and will not adversely affect CV joint performance.

Reassembly

Refer to illustrations 10.10, 10.16, 10.17a and 10.17b

10   Wrap the axleshaft splines with tape to avoid damaging the boot. Slide the small boot clamp and boot onto the axleshaft, then remove the tape **(see illustration)**.

11   Install the inner bearing assembly on the axleshaft with the previously made matchmarks facing outward.

12   Install the snap-ring in the groove. Make sure it's completely seated by pushing on the inner bearing assembly.

13   Fill the outer race and boot with the specified type and quantity of CV joint grease (normally included with the new boot kit). Pack the inner bearing assembly with grease, by hand, until grease is worked completely into the assembly.

14   Slide the outer race down onto the inner race and install the wire ring retainer.

15   Wipe any excess grease from the axle boot groove on the outer race. Seat the small diameter of the boot in the recessed area on the axleshaft. Push the other end of the boot onto the outer race.

16   Equalize the pressure in the boot by inserting a dull screwdriver between the boot and the outer race **(see illustration)**. Don't damage the boot with the tool.

17   Install the boot clamps **(see illustrations)**.

18   Install a new circlip on the inner CV joint stub axle.

19   Install the driveaxle as described in Section 10.

Outer CV joint and boot

Disassembly

Refer to illustration 10.21

20   Following Steps 1 through 8, remove the inner CV joint from the axleshaft.

21   Remove the outer CV joint boot clamps, using the technique described in Step 3. Slide the boot off the axleshaft **(see illustration)**.

Inspection

Refer to illustration 10.23

22   Thoroughly wash the inner and outer CV joints in clean solvent and blow them dry with compressed air, if available. **Note:** *Because the outer joint cannot be disassembled, it is difficult to wash away all the old grease and to rid the bearing of solvent once it's clean. But it is imperative that the job be done thoroughly, so take your time and do it right.*

23   Bend the outer CV joint housing at an angle to the driveaxle to expose the bearings, inner race and cage **(see illustration)**. Inspect the bearing surfaces for signs of wear. If the bearings are damaged or worn, replace the driveaxle.

10.23 After the old grease has been rinsed away and the solvent has been blown out with compressed air, rotate the outer joint housing through its full range of motion and inspect the bearing surfaces for wear and damage - if any of the balls, the race or the cage look damaged, replace the driveaxle and outer joint assembly

11.4 Remove the upper and front mounting bolts, nuts and collars

11.6 . . . then remove the rear mounting bolt and nut (arrow), followed by the front mounting bracket and bolt

11.7 Pull the differential forward to separate the driveshaft from the engine

11.8 Pull out the driveshaft and remove the spring (arrow)

Reassembly

24 Slide the new outer boot onto the driveaxle. It's a good idea to wrap vinyl tape around the spline of shaft to prevent damage to the boot **(see illustration 10.10)**. When the boot is in position, add the specified amount of grease (included in the boot replacement kit) to the outer joint and the boot (pack the joint with as much grease as it will hold and put the rest into the boot). Slide the boot on the rest of the way and install the new clamps **(see illustrations 10.17a and 10.17b)**.
25 Proceed to clean and install the inner CV joint and boot by following Steps 9 through 18, then install the driveaxle as outlined in Section 10.

11 Front differential and driveshaft - removal, inspection and installation

Removal

Refer to illustrations 11.4, 11.6, 11.7 and 11.8

1 Refer to Section 9 and remove one of the front driveaxles.
2 Remove the left front mudguard (see Chapter 7).
3 Disconnect the differential breather hose.

4 Remove the nut from the upper mounting bolt and remove the bolt and collar **(see illustration)**.
5 Remove the forward mounting bolt, then unbolt the mounting bolt bracket from the frame **(see illustration 11.4)**.
6 Remove the rear mounting bolt and nut **(see illustration)**.
7 Pull the differential forward to disengage the driveshaft coupling from the engine **(see illustration)**, then lift out the differential and driveshaft. As you remove the differential, separate the remaining driveaxle from it.
8 Pull the driveshaft coupling off the driveshaft and remove the spring **(see illustration)**. Pull the driveshaft out of the differential.

Inspection

Refer to illustration 11.11

9 Roll the O-rings off the ends of the driveshaft boots, then slide the boots off the shaft. Check the boots for wear, damage and deterioration and replace them if these conditions are found.
10 Check the driveshaft for bending and for worn or damaged splines. Replace it if these conditions are found.
11 Check the oil seals at the driveaxle holes and the driveshaft hole for signs of leakage. Look into the driveaxle holes and check for obvious signs of wear and for damage such as broken gear teeth **(see illustration)**. Turn the pinion (where the driveshaft enters the differen-

11.11 Check the driveshaft oil seal on each side of the differential (arrow) for wear or signs of leakage

12.4 Remove the bolts (arrows) and lower the skid plate away from the differential

12.5 Remove the lower shock absorber bolt and axle housing nuts (arrows) (one nut hidden)

tial) by hand to rotate the gears so they can be inspected and listened to. Slip the driveshaft back in and use it as a handle if necessary.

12   Differential overhaul is a complicated procedure that requires several special tools, for which there are no readily available substitutes. If there's visible wear or damage, or if the differential's rotation is rough or noisy, take it to a Honda dealer for disassembly and further inspection.

Installation

13   Install the boots on the ends of the driveshaft, using new O-rings.
14   Make sure the circlip is in place on the front end of the driveshaft.
15   Lubricate the lips of the driveaxle seals and the driveshaft seal, as well as the splines of the driveshaft, with moly-based multi-purpose grease.
16   Place the differential in the frame, slightly forward of its installed position.
17   Slip the driveshaft into the differential, making sure the circlip engages securely. Be careful not to bend back the rubber boot.
18   Slip the coupling onto the rear end of the driveshaft, making sure to install the spring. Push the coupling inside the rubber boot so the boot's O-ring aligns with the groove in the coupling.
19   Slide the differential back and slip the driveshaft onto the engine shaft. As you reposition the differential, install the driveaxle that was disconnected but not removed from the vehicle.
20   Install the rear mounting nut and bolt and tighten them to the torque listed in this Chapter's Specifications.
21   Install the upper mounting bolt, collar and nut and tighten them to the torque listed in this Chapter's Specifications.
22   Install the front mounting bracket and tighten its bolt securely.
23   Install the front mounting bolt and nut and tighten them to the torque listed in this Chapter's Specifications.
24   The remainder of installation is the reverse of the removal steps.
25   Fill the differential with the recommended oil (see Chapter 1).

12   Rear axle shaft and housing - removal, inspection and installation

Removal

1   Securely block the front wheels so the vehicle won't roll. Loosen the rear wheel nuts with the vehicle on the ground. Jack up the rear end and support it securely, positioning the jackstands so they won't obstruct removal of the axle.
2   Remove the rear wheels and their hubs (see Chapter 6).

350 models
Refer to illustrations 12.4, 12.5, 12.6 and 12.7

3   Remove the brake drum cover, brake drum and brake panel (see Chapter 6). The brake shoes need not be removed from the panel.
4   Remove the skid plate from under the rear differential **(see illustration)**.
5   Remove and discard the nuts that secure the axle housing to the left side of the differential **(see illustration)**. Pull the axle housing off the differential and remove the O-ring.
6   Slide the collar off the left end of the axle shaft **(see illustration)**.
7   Clean any foreign material from the left side of the axle so it won't be pulled into the final drive unit during removal. Tap on the left end of the axle with a soft faced hammer to free it, then pull it out of the axle housing **(see illustration)**.

12.6 Note which way the chamfered side of the collar faces, then slide it off the shaft

12.7 Tap on the left end of the axle shaft (arrow) and pull it out of the differential

12.9a Hold the axle nut with the special tools and unscrew the locknut, then unscrew the axle nut

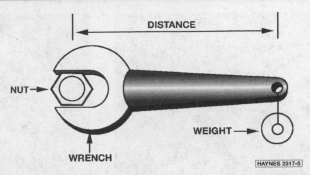

12.9b Length in feet times weight in pounds equals foot-pounds of torque

250 models

Refer to illustrations 12.9a and 12.9b

8   If you're working on a Recon, remove the skid plate from under the rear differential (see illustration 12.4).

9   Unscrew the locknut from the right side of the axle. This requires a 41mm wrench and torque wrench adapter (see illustrations). Similar tools are available from aftermarket suppliers. Because the locknut is tightened to a high torque, whatever tools you use must grip the nuts very securely. The tools, or their equivalents, are necessary to torque the nut and locknut accurately.

10   Unscrew the axle nut once the locknut has been removed.

11   Refer to Chapter 6 and remove the rear brake panel and drum.

12   Tap the axle out of the differential with a soft-faced mallet.

Inspection

13   Check the axle for obvious damage, such as step wear of the splines or bending, and replace it as necessary.

14   Install the wheel hubs on the axle. Place the axle in V-blocks and set up a dial indicator contacting the center of the axle shaft. Rotate the axle and compare runout to the value listed in this Chapter's Specifications. If runout is excessive, replace the axle.

15   Inspect the bearing (inside the axle housing on 350 models and inside the rear portion of the swingarm tube on 250 models). If it's rough, loose or noisy when turned, replace it as described in the following steps.

16   Pry the bearing grease seal out of the axle housing or swingarm tube. If you're working on a 250 model, pry out the stopper ring as well.

17   Insert a long rod through the axle housing or swingarm tube and drive the bearing out towards the seal.

18   Drive in a new bearing, sealed side first, using a bearing driver or socket the same diameter as the bearing outer race. Drive the bearing in to the depth listed in this Chapter's Specifications.

19   If you're working on a 250 model, install the stopper ring.

20   Drive in a new seal, open side first, using a seal driver or block of wood.

Installation

21   Installation is the reverse of removal, with the following additions:

22   Lubricate the axle splines with multipurpose grease.

23   Install the axle from the right side of the vehicle, aligning the splines of the axle with those of the final drive unit.

24   If you're working on a 350 model, coat a new O-ring with grease and install it on the left side of the differential. Install the left axle housing, using new nuts, and tighten them to the torque listed in this Chapter's Specifications.

25   Install the rear brake panel, drum and drum cover (see Chapter 6).

26   If you're working on a 350 model, install the collar on the end of the axle, making sure it faces the proper direction (see illustration 12.6).

27   The remainder of installation is the reverse of the removal steps.

28   Check oil level in the rear differential and add oil as necessary (see Chapter 1).

13   Rear differential - removal, inspection and installation

Removal

Refer to illustrations 13.2a through 13.2e

1   Remove the axle shaft (Section 12).

2   Disconnect the breather tube and remove the bolts that secure the final drive unit to the swingarm (see illustrations). Pull the unit

13.2a On Rancher models, disconnect the breather tube and unbolt the axle housing (arrows) . . .

13.2b . . . disconnect a second breather hose and unbolt the differential from the swingarm

13.2c On Recon/TRX250EX models, disconnect the breather hoses and unbolt the differential from the swingarm at the side and front (two front bolts hidden)

13.2d Pull the final drive unit away from the swingarm and remove the O-ring(s) (arrow) (Rancher shown; Recon/ TRX250EX similar)

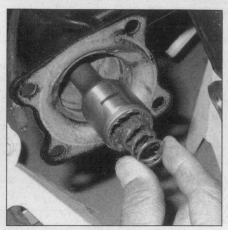

13.2e Remove the spring from the end of the driveshaft

13.3 Rotate the pinion (arrow) and check for rough or noisy movement

rearward to detach it and lift it away from the swingarm, then remove the driveshaft spring and O-rings from the driveshaft **(see illustrations)**.

Inspection

Refer to illustration 13.3

3   Look into the axle holes and check for obvious signs of wear and for damage such as broken gear teeth. Also check the seals for signs of leakage. Turn the pinion by hand (slip the driveshaft back on and use it as a handle if necessary) **(see illustration)**.

4   Differential overhaul is a complicated procedure that requires several special tools, for which there are no readily available substitutes. If there's visible wear or damage, or if the differential's rotation is rough or noisy, take it to a Honda dealer for disassembly and further inspection.

Installation

5   Make sure the driveshaft spring is in place in the driveshaft, then lubricate a new O-ring with oil and install it on the swingarm **(see illustrations 13.2c and 13.2b)**.

6   Align the differential bolts holes with the holes in the swingarm, align the driveshaft with the final drive pinion and push the differential into position.

7   Install the differential bolts and tighten them slightly. Connect the breather hose.

15.8 Loosen the boot clamps so the boot can be detached from the engine as the swingarm is removed

8   Tighten the differential bolts to the torque listed in this Chapter's Specifications.

9   The remainder of installation is the reverse of the removal steps.

14   Swingarm bearings - check

1   Refer to Chapter 6 and remove the rear wheels, then refer to Section 4 and remove the rear shock absorbers.

2   Grasp the rear of the swingarm with one hand and place your other hand at the junction of the swingarm and the frame. Try to move the rear of the swingarm from side-to-side. Any wear (play) in the bearings should be felt as movement between the swingarm and the frame at the front. The swingarm will actually be felt to move forward and backward at the front (not from side-to-side). If any play is noted, the bearings should be replaced with new ones (see Section 16).

3   Next, move the swingarm up and down through its full travel. It should move freely, without any binding or rough spots. If it does not move freely, refer to Section 16 for servicing procedures.

15   Swingarm - removal and installation

1   If the swingarm is being removed just for bearing replacement or driveshaft removal, the brake panel, differential and rear axle need not be removed from the swingarm. **Note:** *Loosening and tightening the pivot bolt locknut on the right side of the swingarm requires a special wrench for which there is no alternative. If you don't have the special Honda tool or an exact equivalent, the locknut must be unscrewed (and later tightened) by a Honda dealer.*

Removal

Refer to illustrations 15.8, 15.9, 15.10, 15.11, 15.12a, 15.12b, 15.2c and 15.12d

2   Raise the rear end of the vehicle off the ground with a jack. Support the vehicle securely so it can't be knocked over while it's jacked up.

3   Remove the mudguards from the lower front edges of the rear fender (see Chapter 7).

4   Remove the rear wheels (see Chapter 6).

5   Refer to Section 4 and unbolt the lower end of the shock absorber from the swingarm.

6   Disconnect both rear brake cables (see Chapter 6).

7   If you're planning to remove the rear axle, differential or brake panel, do it now (see Section 12, Section 13 or Chapter 6).

8   At the front of the swingarm, loosen the clamps that secure the rubber boot **(see illustrations)**.

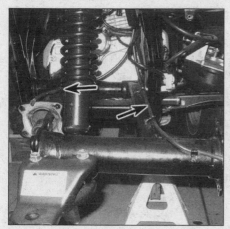

15.9 Detach the breather hoses (arrows) from the top of the swingarm

15.10 Pry the plastic cap from each side of the swingarm

15.11 Unscrew the pivot bolt with a 17mm Allen bolt bit

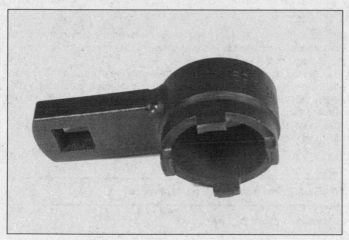

15.12a This tool is used to loosen the pivot bolt locknut as well as to tighten it to the correct torque

15.12b Insert the Allen bolt bit in the right pivot bolt . . .

9   Free the breather tubes from their retainer on top of the swingarm **(see illustration)**.
10   Pry the plastic pivot cap from each side of the swingarm **(see illustration)**.
11   Unscrew the pivot bolt from the left side of the vehicle with a

socket and an Allen bolt bit **(see illustration)**. These are available from tool stores.
12   On the right side of the vehicle, prevent the pivot bolt from turning with the Allen bolt bit and unscrew the pivot bolt locknut with the special Honda tool (tool number 07908-4690003) **(see illustrations)**. Once

15.12c . . . and place the special tool on the locknut; loosen the locknut with the special tool . . .

15.12d . . . then unscrew the locknut and pivot bolt

16.3 Pry the grease seals from the swingarm . . .

16.4 . . . to expose the bearings for inspection

the locknut is loose, unscrew the pivot bolt.

13   Pull the swingarm back and away from the vehicle, separating the driveshaft from the engine as you pull.

14   Check the pivot bearings in the swingarm for dryness or deterioration (see Section 16). If they're in need of lubrication or replacement, refer to Section 16.

Installation

15   If the driveshaft was removed from the swingarm, install it. Lubricate the driveshaft splines with molybdenum disulfide grease.

16   If the boot was removed from the swingarm, install it with the clamp at the engine end upward **(see illustration 15.8)**.

17   Lift the swingarm into position in the frame. Align the splines of the driveshaft with those of the engine and align the pivot bolt holes in the swingarm with those in the frame.

18   Install the pivot bolt in the left side of the swingarm. Use the Allen bolt bit to tighten it to the torque listed in this Chapter's Specifications.

19   Install the pivot bolt in the right side of the swingarm and tighten it to the torque listed in this Chapter's Specifications.

20   Raise and lower the swingarm several times, moving it through its full travel to seat the bearings and pivot bolts.

21   Retighten the pivot bolt in the right side of the swingarm to the torque listed in this Chapter's Specifications.

22   Hold the pivot bolt from turning with the Allen bolt bit and use the locknut wrench to tighten the locknut to the torque listed in this Chap-

ter's Specifications. **Note:** *The reading shown on the torque wrench is lower than the actual locknut torque because the special tool increases the leverage of the torque wrench.*

23   Slip the boot onto the engine. Make sure the clamp on the engine end is upward, with the screw head facing in the proper direction **(see illustration 15.8)**. Position the screws on the boot clamps so they align with the tabs on the boot. The forward clamp screw aligns with the outer boot tab. The rear clamp screw aligns with the upper boot tab. Tighten the clamps.

24   The remainder of installation is the reverse of the removal steps.

16   Swingarm bearings - replacement

Refer to illustrations 16.3 and 16.4

1   The swingarm pivot shafts ride on ball bearings.

2   Remove the swingarm (see Section 15).

3   Pry the seal from each side of the swingarm **(see illustration)**.

4   Rotate the center race of each ball bearing and check for roughness, looseness or play **(see illustration)**. If there's any doubt about bearing condition, replace the bearings as a set.

5   Bearing replacement requires a slide hammer and blind hole puller, which can be rented from tool rental dealers. Pull the old bearings out, using the special tools. Drive in the bearing grease holder, then drive in the bearing, using a bearing driver or socket.

6   Pack the grease retainer inside each bearing, and the space between the bearing balls, with molybdenum disulfide grease.

7   If the bearings are replaced, the inner races must also be replaced. Inner race replacement requires a slide hammer and blind hole puller, which can be rented from tool rental dealers. Pull the old races out, using the special tools. Drive in the new races with a bearing driver or socket.

8   Pack the bearings and coat the seal lips with molybdenum disulfide grease.

17   Rear driveshaft - removal, inspection and installation

Removal

Refer to illustration 17.2

1   Refer to Section 16 and remove the swingarm.

2   Pull the driveshaft out of the swingarm, then pull the spring out of the driveshaft **(see illustration 11.8 and the accompanying illustration)**.

17.2 Pull the driveshaft out of the swingarm . . .

Inspection

Refer to illustration 17.4

3   Check the shaft for bending or other visible damage such as step wear of the splines. If the shaft is bent, replace it. If the splines at the rear end of the shaft are worn, replace the shaft. If the splines at the front end of the shaft (the universal joint) are worn, replace the universal joint.

4   Hold the driveshaft firmly in one hand and try to twist the universal joint **(see illustration)**. If there's play in the joint, replace it (but don't confuse play in the joint with its normal motion). Parts are not available to overhaul the U-joint, but it can be replaced separately from the driveshaft.

Installation

5   Installation is the reverse of the removal procedure. Lubricate the driveshaft splines with molybdenum disulfide grease. Don't forget to reinstall the spring.

17.4 If the universal joint is loose or clicks when it rotates, replace it

Notes

Chapter 6
Brakes, wheels and tires

Contents

	Section
Brake fluid level check and fluid change	See Chapter 1
Brake hoses and lines - inspection and replacement	9
Brake lining wear check and system check	See Chapter 1
Brake pedal, rear brake lever and cables - removal and installation	13
Brake system bleeding	8
Front brake caliper (TRX250EX models) - removal, overhaul and installation	15
Front brake disc and hub (TRX250EX) - inspection, removal and installation	16
Front brake drums - inspection, waterproof seal replacement and 2WD bearing replacement	3
Front brake drums - removal and installation	2
Front brake master cylinder - removal, overhaul and installation	7

	Section
Front brake pads (TRX250EX models) - replacement	14
Front brake panel - removal, inspection and installation	6
Front brake shoes - removal, inspection and installation	4
Front wheel bearings (TRX250EX models) - removal, inspection and installation	17
Front wheel cylinders - removal, overhaul and installation	5
General information	1
Rear brake drum - removal, inspection and installation	10
Rear brake panel - removal, inspection and installation	12
Rear brake shoes - removal, inspection and installation	11
Rear wheel hubs - removal and installation	19
Tires - general information	20
Tires and wheels - general check	See Chapter 1
Wheels - inspection, removal and installation	18

Specifications

Brakes

Brake fluid type	See Chapter 1
Brake lining minimum thickness	See Chapter 1
Brake pedal height	See Chapter 1
Front drum diameter	
Rancher models	
Standard	160 mm (6.299 inches)
Wear limit	161 mm (6.338 inches)*
Recon models	
Standard	130 mm (5.12 inches)
Wear limit	131 mm (5.16 inches)
Rear drum diameter	
Rancher models	
Standard	160 mm (6.299 inches)
Wear limit	161 mm (6.338 inches)*
Recon and TRX250EX models	
Standard	140 mm (5.51 inch)
Wear limit	141 mm (5.55 inch)
Front disc (TRX250EX models)*	
Thickness	
Standard	2.8 to 3.2 mm (0.11 to 0.13 inch)
Limit	2.5 mm (0.10 inch)
Front waterproof seal lubricant	
Type	NLGI no. 3
Amount	14 to 16 grams (0.5 to 0.6 oz)

Refer to marks cast into the drum or disc (they supersede information printed here)

Wheels and tires

Tire pressures	See Chapter 1
Tire tread depth..	See Chapter 1
Torque specifications	
Drum to hub bolts (Rancher models)	Not specified
Front driveaxle (hub) nuts	
Drum brake models	78 Nm (58 ft-lbs)
Disc brake models	69 Nm (51 ft-lbs)
Wheel cylinder bolts/nuts (front drum brakes)	8 Nm (72 inch-lbs)
Caliper mounting bolts (front disc brakes)	30 Nm (22 ft-lbs)*
Pad pins (front disc brakes)...	18 Nm (13 ft-lbs)
Pad pin plugs (front disc brakes)...	3 Nm (26 inch-lbs)
Disc-to-hub bolts (front disc brakes)..	42 Nm (31 ft-lbs)*
Front brake hose union bolts..	34 Nm (25 ft-lbs)
Front brake pipe to wheel cylinder joint nuts (Rancher models)	16 Nm (144 inch-lbs)
Master cylinder cover screws	
Drum brake models	2 Nm (17 inch-lbs)
Disc brake models	1.5 Nm (13 inch-lbs)
Master cylinder clamp bolts	
Drum brake models	12 Nm (108 inch-lbs)
Disc brake models	Not specified
Brake panel bolts/nuts (drum brake models)*	
Front brake..	29 Nm (22 ft-lbs)
Rear brake	
Rancher	35 Nm (25 ft-lbs)
Recon and TRX250EX	Not specified
Rear brake panel drain bolt..	12 Nm (84 inch-lbs)
Rear wheel hub nuts	
Rancher and Recon models	137 Nm (101 ft-lbs)
TRX250EX models	
2005 and earlier..	147 Nm (108 ft-lbs)
2006 and later...	137 Nm (101 ft-lbs)

Discard the bolts or nuts and replace them with new ones each time they're removed.

---

1   General information

Rancher and Recon models are equipped with hydraulic drum brakes on the front wheels. TRX250EX models use hydraulic disc brakes at the front. All models use a single mechanical drum brake mounted on the rear axle inboard of the rear wheels. All drum brakes are sealed to keep water out. The front brakes are actuated by a lever on the right handlebar. The rear brakes have two means of actuation: a lever on the left handlebar, which provides a parking brake, and a pedal on the right side of the vehicle. The pedal and lever are connected to the rear brake assembly by cables.

All models are equipped with steel wheels, which require very little maintenance and allow tubeless tires to be used. **Caution:** *Brake components rarely require disassembly. Do not disassemble components unless absolutely necessary. If any brake line connection in the system is loosened, the entire system should be disassembled, drained, cleaned and then properly filled and bled upon reassembly. Do not use solvents on internal hydraulic brake components. Solvents will cause seals to swell and distort. Use only clean brake fluid for cleaning. Use care when working with brake fluid as it can injure your eyes and it will damage painted surfaces and plastic parts.*

2   Front brake drums - removal and installation

Warning: *If a front wheel cylinder indicates the need for an overhaul (usually due to leaking fluid or sticky operation), ALL FOUR front wheel cylinders (Rancher models) or BOTH front calipers (Recon models) should be overhauled and all old brake fluid flushed from the system. Also, the dust created by the brake system is harmful to your health. Never blow it out with compressed air and don't inhale any of it. An approved filtering mask should be worn when working on the brakes.*

Do not, under any circumstances, use petroleum-based solvents to clean brake parts. Use clean brake fluid, brake cleaner or denatured alcohol only!

Removal

Refer to illustrations 2.3 and 2.5

1   Loosen the front wheel nuts. Securely block the rear wheels so the vehicle can't roll. Jack up the front end and support it securely on jackstands.

2   Refer to Section 15 and remove the front wheel.

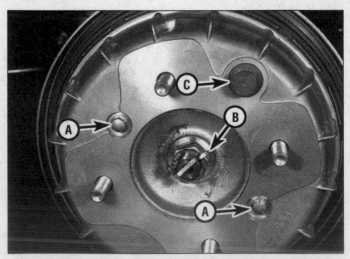

2.3 Loosen the drum-to-hub bolts (A) if you plan to separate the drum and hub; bend back the cotter pin (B) and remove the nut; the rubber plug (C) is for front brake adjustment

2.5 Replace the O-ring if the drum is separated from the hub

2.6 Pack grease into the space between the seal lips (the correct amount is listed in this Chapter's Specifications); be careful not to get any inside the drum

3   If you're going to remove the front hub from a Rancher, remove the cotter pin from the front hub nut **(see illustration)**. Have an assistant hold the brake on and remove the nut with a socket and breaker bar. This is easier to do now, since you can use the front brakes to keep the hub from turning while you loosen the nut. If you're working on a Recon, the cotter pin and front hub nut must be removed to get the brake drum off.

4   If you're working on a Rancher, remove the drum-to-hub bolts **(see illustration 2.3)**. Pull the brake drum off.

5   If you're working on a Rancher, inspect the hub O-ring **(see illustration)**. Since the O-ring's purpose is to keep water out of the brakes, replace it if its condition is in doubt.

Installation

Refer to illustration 2.6

6   Installation is the reverse of the removal steps, with the following additions:

a) *Pack the space between the drum seal lips with multipurpose grease* **(see illustration)**. *Be sure not to get any grease on the inside of the drum; if you do, clean it off with a non-residue solvent such as brake cleaner or lacquer thinner.*

b) *Tighten the drum-to-hub bolts (Rancher only) to the torque listed in this Chapter's Specifications.*

c) *On Recon models, be sure to install the collar in the dust seal.*

d) *Tighten the hub nut if it was removed to the torque listed in this Chapter's Specifications and install a new cotter pin. If necessary, tighten the nut to align the hole in the spindle with the slots in the nut.*

e) *Once the nut is tightened properly, bend the cotter pin to secure it.*

f) *Refer to Chapter 1 and adjust the brakes.*

3   Front brake drums - inspection, waterproof seal replacement and 2WD bearing replacement

Inspection

Refer to illustration 3.1

1   Check the brake drum for wear or damage. Measure the diameter at several points with a drum micrometer (or have this done by a dealer service department or other qualified shop). If the measurements are uneven (indicating that the drum is out-of-round) or if there are scratches deep enough to snag a fingernail, replace the drum. The drum must also be replaced if the diameter is greater than that cast inside the drum **(see illustration)**. Honda recommends against turning brake drums.

2   Check the waterproof seal on the edge of the brake drum for wear (caused by rubbing against the brake panel). To do this, measure the

length of the seal lip from the point where it contacts the brake drum to the point where it contacts the brake panel **(see illustration 3.1)**. Also check for damage such as cuts and tears. If the seal on a Rancher is worn or damaged, replace it as described below. The seal on Recon models is not available separately from the brake drum.

Waterproof seal replacement (Rancher)

Refer to illustration 3.7

3   This procedure is somewhat complicated and must be done correctly to ensure a watertight seal between the brake drum and panel. It requires a hydraulic press and a press plate bigger than the brake drum. An additional steel plate 140 mm (5.5 inches) in diameter and at least 10 mm (0.4 inch) thick is required to support the brake drum so it isn't warped during the press operation. If you don't have the necessary equipment, have this procedure done by a Honda dealer.

4   Carefully pry the waterproof seal off the edge of the brake drum.

5   It's important to press the new seal onto the drum just far enough, but not too far. To make sure this happens, you'll need to calculate the final clearance between the outer circumference of the drum and the seal, as well as between the inner circumference of the drum and the seal. These clearances will become smaller as the seal is pressed onto the drum, so calculating the clearances in advance will let you know when the seal has been pressed on far enough.

6   To determine what the final clearances should be, measure as follows.

3.1 The maximum diameter (A) is cast inside the brake drum; measure the length of the seal from the lip (B) to the point where it contacts the shoulder on the outer circumference of the brake drum

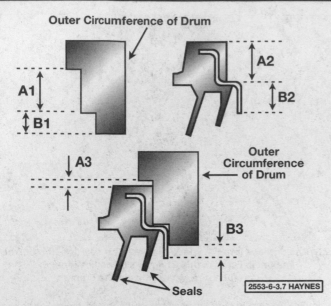

3.7 Calculate the seal's installed clearances - subtract measurement A2 from A1 to get clearance A3; subtract measurement B2 from B1 to get clearance B3

7   Measure the depth of both recesses in the brake drum and the seal surfaces that will contact them **(see illustration)**. Calculate the difference between these measurements to get the final clearances for the seal as follows:

a) *Subtract measurement A2 from measurement A1 to get clearance A3.*

b) *Subtract measurement B2 from Measurement B1 to get clearance B3.*

8   Write the clearances down for later use.

9   Dip the new seal in water to lubricate it (don't use any other type of lubricant). Place the new seal on the edge of the brake drum, then set the drum seal side down on a press plate.

10   Place a steel plate about 5.5 mm (1.6 inch) in diameter and at least 10 mm (0.4 inch) thick on top of the brake drum so it won't collapse from the force of the press.

11   Slowly and carefully press the brake drum down into the seal, making sure not to press it too far. If the seal is damaged or is pressed on too far, remove it and start over with a new seal. Stop pressing when the measured clearances are equal to those written down in Step 7.

Recon wheel bearing replacement

12   The front wheel bearings on Recon models are mounted inside the brake drums.

13   Pry out the seal.

14   Insert a soft metal drift into the hub from the inner side and rest it against the outer bearing. Tap gently against the drift, moving it back and forth to opposite sides of the bearing, to drive the bearing from the drum. Then insert the bearing from the other side of the drum and tap the inner bearing out in the same way.

15   Pack the new bearings with multi-purpose grease. Work the grease into the spaces between the bearing balls. Hold the outer race and rotate the inner race as you pack the bearing to distribute the grease.

16   Place the new outer bearing in the brake drum with its sealed side out. Position the drum seal side down on a workbench or similar surface. Support the center of the drum from below so the drum isn't resting on the waterproof seal. Tap the bearing into position with a bearing driver or socket the same diameter as the bearing outer race. Install a new seal, with its open side toward the bearing. Place a seal driver, large socket or block of wood against the seal and tap it into position.

17   Turn the drum over and install the inner bearing and seal in the same manner.

4   Front brake shoes - removal, inspection and installation

Removal

Refer to illustrations 4.2 and 4.3

1   Refer to Section 2 and remove the brake drum.

2   Rotate the retainer pins with pliers to align them with the slots in the pin holders, then remove the pin holders **(see illustration)**.

3   Pull the shoes apart against the force of the spring(s) and take them off the brake panel **(see illustration)**.

Inspection

Refer to illustration 4.8

4   Check the linings for wear, damage and signs of contamination from road dirt or water. If the linings are visibly defective, replace them.

5   Measure the thickness of the lining material (just the lining material, not the metal backing) and compare with the value listed in the Chapter 1 Specifications. Replace the shoes if the material is worn to the minimum or less.

6   Check the ends of the shoes where they contact the wheel cylinders. Replace the shoes if there's visible wear.

7   Pull back the wheel cylinder cups and check for fluid leakage. Slight moisture inside the cups is normal, but if fluid runs out, refer to Section 5 and replace the wheel cylinder(s).

8   Inspect the rubber seals on the retaining pins **(see illustration)**. Replace them if they're worn, hardened or deteriorated. The seals are meant to keep out water, not just dust, so replace them if there's any doubt about their condition.

4.2 Rotate the ends of the retainer pins to align with the slot in the pin holder, then take the pin holders off

4.3 Pull the shoes apart and take them off the brake panel

4.8 Inspect the rubber grommets on the retainer pins and replace them if they're damaged or deteriorated

4.11a The assembled right brake should look like this (right side) . . .

4.11b . . . and the assembled left brake should look like this (Rancher shown; Recon similar)

4.12 Compress the pin holders and turn the pins 90-degrees so they hold the pins securely

Installation

Refer to illustrations 4.11a, 4.11b and 4.12

9   Apply high temperature grease to the ends of the springs and shoes. Apply a thin smear of grease to each of the brake shoe contact points on the brake panel.

10   Hook the springs to the shoes, noting how they're installed (the hooked ends of the springs face outward).

11   Pull the shoes apart and position their ends in the wheel cylinders **(see illustrations)**. The flatter ends of the brake shoes fit into the wheel cylinder(s); the other ends fit into the adjuster(s).

12   Install the pin holders and pins. Compress the holders and turn the pins 90-degrees so the pins secure the holders **(see illustration)**.

13   The remainder of installation is the reverse of the removal steps.

5   Front wheel cylinders - removal, overhaul and installation

Removal

Refer to illustrations 5.3a, 5.3b and 5.3c

1   Refer to Sections 2 and 4 and remove the brake drum and shoes.

2   Rancher models use two wheel cylinders which incorporate adjuster mechanisms **(see illustrations 4.11a and 4.11b)**. Recon models use a single wheel cylinder between the upper ends of the shoes and an adjuster between the lower ends of the shoes.

3   From the back side of the brake panel, detach the brake hose

5.3a Unscrew the union bolt; use a new sealing washer on each side of the bolt on installation and position the neck of the hose between the stoppers (arrow)

from the wheel cylinder(s) **(see illustration)**. Remove the union bolt and sealing washers. On Rancher models, unscrew the brake pipe nuts **(see illustrations)**. Remove the wheel cylinder nuts and bolts (Rancher) or bolts (Recon). Work the wheel cylinder free of the sealant that secures it to the brake panel and take it off. If you're working on a Recon, unbolt the adjuster body and remove it from the brake panel.

5.3b Unscrew the brake pipe fitting (arrow) from the wheel cylinder with a flare nut wrench; remove the bolt and the nut (B)

5.3c Loosen the other end of the brake pipe and remove the nut and bolt

5.4a Wheel cylinder components

1	Boot	4	Wheel cylinder	7	Adjuster wheel
2	Piston		body	8	Adjuster screw
3	Piston cup seal	5	Lock spring		(left-hand
		6	Screw		threads)

Overhaul

Refer to illustrations 5.4a and 5.4b

4   Remove the boot(s) from the cylinder(s) **(see illustration)**. If you're working on a Rancher, remove the adjuster lock spring and pull the adjuster out of the other end of the cylinder **(see illustration)**. If you're working on a Recon, remove the piston clips.

5   Push the piston(s) out of the cylinder.

6   Check the piston and cylinder bore for wear, scratches and corrosion. If there's any doubt about their condition, replace the cylinder as an assembly. Even barely visible flaws can reduce braking performance.

7   The piston cups and boots are available separately and should be replaced whenever the wheel cylinders are overhauled. Work the cup(s) off the piston(s). Dip new ones in clean brake fluid and carefully install them without stretching or damaging them. The wide side of the piston cup faces into the cylinder bore.

8   Remove the screw from the adjuster nut. Check the adjuster components for wear or damage and replace as necessary.

9   Assembly is the reverse of the disassembly steps, with the following additions:

a) *Coat the cylinder bore with clean brake fluid. Install the piston with the wide side of the piston cup entering the bore first. Be sure not to turn back the lip of the cup.*

b) *Lubricate the adjuster wheel(s) with silicone grease.*

Installation

Refer to illustration 5.10

10   Installation is the reverse of the removal steps, with the following additions:

a) *Apply a thin coat of sealant to the wheel cylinder contact area on the brake panel **(see illustration)**.*

b) *Be sure to install the wheel cylinders on the correct sides of the vehicle. Right wheel cylinders have the letter R cast into the cylinders body; left wheel cylinders are marked with an L.*

c) *Tighten the wheel cylinder bolts and nuts to the torque listed in this Chapter's Specifications.*

d) *Use new sealing washers on the brake hose union bolt. Position the neck of the brake hose between the stoppers on the brake panel. Tighten the union bolt and the metal brake line fittings to the torque listed in this Chapter's Specifications.*

e) *Refer to Section 8 and bleed the brakes.*

6   Front brake panel - removal, inspection and installation

Removal

Refer to illustrations 6.2, 6.3 and 6.4

1   Refer to Section 2 and remove the brake drum. If you're planning to remove the brake shoes, refer to Section 4 (the brake panel can be removed with the shoes installed).

2   Disconnect the breather hose and remove the brake hose union bolt **(see illustration 5.2 and the accompanying illustration)**. If you're planning to remove the wheel cylinders, this is a good time to loosen the bolts and nuts, while the brake panel is securely bolted to the knuckle. The brake panel can be removed with the wheel cylinders installed.

3   Remove the brake panel mounting bolts **(see illustration)**. Lift the brake panel off the knuckle. Note: Discard the bolts and use new ones on installation. The bolt threads have a special dry waterproof coating.

4   If you're working on a Rancher, remove the O-ring from the knuckle and install a new one **(see illustration)**.

5   Installation is the reverse of the removal steps, with the following additions:

a) *Use new brake panel bolts and tighten them to the torque listed in this Chapter's Specifications.*

b) *Use new sealing washers on the brake hose union bolt (if equipped). Position the neck of the bolt between the stoppers on the brake panel and tighten the bolt to the torque listed in this Chapter's Specifications.*

5.4b Remove the screw and take off the lock spring

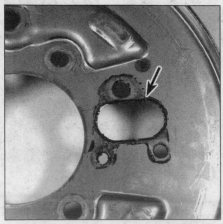

5.10 Apply sealant to the wheel cylinder contact areas

6.2 Disconnect the breather hose

6.3 Remove the brake panel mounting bolts (arrows)

6.4 Remove the O-ring (arrow) from the knuckle

7.4 Remove the union bolt (arrow); on installation, use a new sealing washer on each side of the bolt

7   Front brake master cylinder - removal, overhaul and installation

1   If the master cylinder is leaking fluid, or if the lever doesn't produce a firm feel when the brake is applied, and bleeding the brakes does not help, master cylinder overhaul is recommended. Before disassembling the master cylinder, read through the entire procedure and make sure that you have the correct rebuild kit. Also, you will need some new, clean brake fluid of the recommended type, some clean rags and internal snap-ring pliers. **Note:** To prevent damage to the paint from spilled brake fluid, always cover the fuel tank when working on the master cylinder.

2   **Caution:** *Disassembly, overhaul and reassembly of the brake master cylinder must be done in a spotlessly clean work area to avoid contamination and possible failure of the brake hydraulic system components.*

Removal

Refer to illustrations 7.4 and 7.5

3   Loosen, but do not remove, the screws holding the reservoir cover in place.

4   Place rags beneath the master cylinder to protect the paint in case of brake fluid spills. Remove the union bolt **(see illustration)** and separate the brake hose from the master cylinder. Wrap the end of the hose in a clean rag and suspend the hose in an upright position or

bend it down carefully and place the open end in a clean container. The objective is to prevent excess loss of brake fluid, fluid spills and system contamination.

5   Remove the master cylinder mounting bolts **(see illustration)** and separate the master cylinder from the handlebar.

Overhaul

Refer to illustration 7.8

6   Remove the locknut from the underside of the lever pivot screw, then remove the screw.

7   Carefully remove the rubber dust boot from the end of the piston. Remove the separator from the bottom of the reservoir.

8   Using snap-ring pliers, remove the snap-ring **(see illustration)** and slide out the piston, the cup seals and the spring. Lay the parts out in the proper order to prevent confusion during reassembly.

9   Clean all of the parts with brake system cleaner (available at auto parts stores), isopropyl alcohol or clean brake fluid. **Caution:** *Do not, under any circumstances, use a petroleum-based solvent to clean brake parts.* If compressed air is available, use it to dry the parts thoroughly (make sure it's filtered and unlubricated). Check the master cylinder bore for corrosion, scratches, nicks and score marks. If damage is evident, the master cylinder must be replaced with a new one. If the master cylinder is in poor condition, then the wheel cylinders should be checked as well.

10   Remove the old cup seals and install the new ones. One cup seal

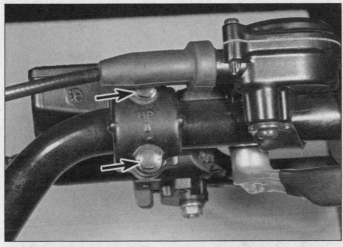

7.5 Remove the mounting bolts (arrows); the UP mark on the clamp must be upright when the clamp is installed

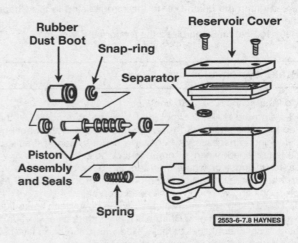

7.8 Master cylinder - exploded view

8.5a Connect a plastic or rubber hose to the bleed valve; open and close the valve with a wrench (this is a drum brake)

8.5b Here's the caliper bleed valve on a TRX250EX

is installed on the piston and the other cup seal is mounted on the end of the spring. Make sure the lips of the cup seals face away from the lever end of the piston. If a new piston is included in the rebuild kit, use it regardless of the condition of the old one.

11   Before reassembling the master cylinder, soak the piston and the rubber cup seals in clean brake fluid for ten or fifteen minutes. Lubricate the master cylinder bore with clean brake fluid, then carefully insert the piston and related parts in the reverse order of disassembly. Make sure the lips on the cup seals do not turn inside out when they are slipped into the bore.

12   Depress the piston, then install the snap-ring (make sure the snap-ring is properly seated in the groove with the sharp edge facing out). Install the rubber dust boot (make sure the lip is seated properly in the piston groove).

13   Install the brake lever and tighten the pivot bolt locknut.

Installation

14   Attach the master cylinder to the handlebar. Align the corner of the master cylinder's front mating surface with the punch mark on the handlebar.

15   Make sure the arrow and the word UP on the master cylinder clamp are pointing up, then tighten the bolts to the torque listed in this Chapter's Specifications. Tighten the top bolt fully, then tighten the lower bolt. **Caution:** *Don't try to close the gap at the lower bolt mating surface or the clamp may break.*

16   Connect the brake hose to the master cylinder, using new sealing washers. Tighten the union bolt to the torque listed in this Chapter's Specifications.

17   Refer to Section 8 and bleed the air from the system.

8   Brake system bleeding

Refer to illustrations 8.5a, 8.5b and 8.5c

1   Bleeding the brake is simply the process of removing all the air bubbles from the brake fluid reservoir, the lines and the wheel cylinders. Bleeding is necessary whenever a brake system hydraulic connection is loosened, when a component or hose is replaced, or when the master cylinder or wheel cylinders are overhauled. Leaks in the system may also allow air to enter, but leaking brake fluid will reveal their presence and warn you of the need for repair.

2   To bleed the brake, you will need some new, clean brake fluid of the recommended type (see Chapter 1), a length of clear vinyl or plastic tubing, a small container partially filled with clean brake fluid, some rags and a wrench to fit the brake caliper bleed valve.

3   Cover the fuel tank and other painted components to prevent damage in the event that brake fluid is spilled.

4   Remove the reservoir cap and slowly pump the brake lever a few times, until no air bubbles can be seen floating up from the holes at the bottom of the reservoir. Doing this bleeds the air from the master cylinder end of the line. Reinstall the reservoir cap.

5   Attach one end of the clear vinyl or plastic tubing to the brake caliper bleed valve and submerge the other end in the brake fluid in the container **(see illustrations)**.

6   Check the fluid level in the reservoir. Do not allow the fluid level to drop below the lower mark during the bleeding process.

7   Carefully pump the brake lever three or four times and hold it while opening the caliper bleed valve. When the valve is opened, brake fluid will flow out of the wheel cylinder into the clear tubing and the lever will move toward the handlebar.

8   Retighten the bleed valve, then release the brake lever gradually. Repeat the process until no air bubbles are visible in the brake fluid leaving the wheel cylinder and the lever is firm when applied. Remember to add fluid to the reservoir as the level drops. Use only new, clean brake fluid of the recommended type. Never reuse the fluid lost during bleeding.

9   Repeat this procedure to the other wheel. Be sure to check the fluid level in the master cylinder reservoir frequently.

10   Replace the reservoir cap, wipe up any spilled brake fluid and check the entire system for leaks. **Note:** *If bleeding is difficult, it may be necessary to let the brake fluid in the system stabilize for a few hours (it may be aerated). Repeat the bleeding procedure when the tiny bubbles in the system have floated out.*

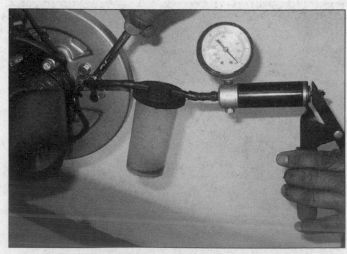

8.5c You can also use a vacuum pump like this one

9.2 Check the hoses for cracks; pay special attention to the points where they meet the metal fittings (arrows)

9.4 Remove the retainers to release the hose

9   Brake hoses and lines - inspection and replacement

Inspection

Refer to illustration 9.2

1   Once a week, or if the vehicle is used less frequently, before every use, check the condition of the brake hoses.

2   Twist and flex the rubber hoses while looking for cracks, bulges and seeping fluid. Check extra carefully around the areas where the hoses connect with metal fittings, as these are common areas for hose failure **(see illustration)**.

Replacement

Refer to illustration 9.4

3   There are two brake hoses. One hose is attached to the master cylinder. The master cylinder hose runs to the metal joint. The metal joint is part of the lower hose assembly, consisting of the joint and two permanently attached lengths of hose which run to the wheel cylinders.

4   Cover the surrounding area with plenty of rags and unscrew the union bolts on either end of the hose. Detach the hose from any retainers that may be present and remove the hose **(see illustration)**.

5   Position the new hose, making sure it isn't twisted or otherwise strained, between the two components. Make sure the metal tube portion of the banjo fitting (if equipped) is located between the stoppers on the component it's connected to. Install the union bolts, using new sealing washers on both sides of the fittings, and tighten them to the torque listed in this Chapter's Specifications. If the hose is connected by a flare nut, hold it with one wrench and tighten the flare nut with another wrench.

6   Flush the old brake fluid from the system, refill the system with the recommended fluid (see Chapter 1) and bleed the air from the system (see Section 8). Check the operation of the brakes carefully before riding the vehicle.

10   Rear brake drum - removal, inspection and installation

Removal

1   Refer to Section 18 and remove the right rear wheel.
2   Refer to Section 19 and remove the right rear wheel hub.

350 models

Refer to illustrations 10.3 and 10.4

3   Remove the drum cover bolts and take the cover off **(see illustration)**.

4   Remove the drum cover O-ring and pull the drum off **(see illustration)**.

250 models

Refer to illustrations 10.6a, 10.6b, 10.7, 10.8a, 10.8b, 10.9a and 10.9b

5   Remove the wheel hub (see Section 19) and the rear axle nuts (see Chapter 5).

6   Remove the lockwasher and O-ring **(see illustrations)**.

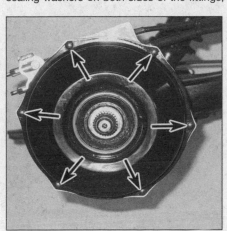

10.3 Remove the bolts (arrows) and take the cover off the brake drum

10.4 Inspect the O-ring (arrow) and replace it if there's any doubt about its condition

10.6a Remove the washer, noting the OUTSIDE mark, and the thrust washer behind it . . .

10.6b . . . and remove the O-ring, using a pointed tool if necessary

10.7 Unscrew the cable wingnuts (arrows) and free the cables from the levers

10.8a Remove the brake panel bolts (arrows)

7　Unscrew the wing nuts from the parking brake and brake pedal cables **(see illustration)**. Label the cables and detach them from the lever, then from the brackets on the brake panel.

8　Remove the brake panel bolts, then carefully pry the panel away from the drum housing, using the pry points **(see illustrations)**.

9　Pull the brake panel off the axle shaft and remove the collar **(see illustration)**. Slide the brake drum off the axle shaft **(see illustration)**.

Inspection

Refer to illustrations 10.10, 10.11 and 10.12

10　Check the drum cover O-ring (350 models) or brake panel O-ring (250 models) for wear, deterioration or hardening **(see illustration 10.4 or the accompanying illustration)**. Since the O-ring is intended to waterproof the rear brake, replace it if there's any doubt about its condition.

10.8b Pry the panel off, using the pry points (arrows); don't pry against the mating surfaces

10.9a Remove the collar

10.9b Once the panel has been removed, take the brake drum off the axle

10.10 Replace the brake panel O-ring (250 models) if its condition is in doubt

10.11 Replace the seal in the center of the drum cover (350 models) if it shows signs of leakage

11.2 Straighten the cotter pins (arrows) and pull them out, then remove the washers

11 If you're working on a 350 model, inspect the seal in the brake drum cover for wear or signs of leakage **(see illustration)**. If any defects are visible, pry the seal out and tap in a new one with a hammer and seal driver or a socket the same diameter as the seal.

12 Refer to Section 11 and look for leaks around the axle seal in the center of the brake panel. Replace the seal as described in Section 12 if there are any signs of leakage. If you're working on a 250 model, look for leaks in the axle seal in the right side of the differential.

13 Refer to Section 3 for drum inspection details.

Installation

14 Installation is the reverse of the removal steps. Coat the lip of the drum cover seal with grease.

---

11 Rear brake shoes - removal, inspection and installation

Removal

Refer to illustration 11.2

1 Refer to Section 10 and remove the rear brake drum.

2 If you're planning to reinstall the brake shoes, mark them so they can go in their original locations. If you're working on a 350 model, straighten and pull out the cotter pins that secure the brake shoe washer and remove the washer **(see illustration)**.

3 Unhook and remove the brake springs. Slide the shoes off the pivot pins.

Inspection

Refer to illustration 11.5

4 Refer to Section 4 for shoe and lining inspection details.

5 Check the brake cam and pivot pin(s) for wear or damage **(see illustration)**. If the brake cam shows wear or damage, refer to Section 12 and remove it from the brake panel. The pivot pin(s) are permanently attached to the brake panel, so the panel will have to be replaced if they're worn.

Installation

6 Installation is the reverse of the removal steps, with the following additions:

   a) *Apply a thin film of high-temperature brake grease to the brake cam and pivot pin(s), as well as to the shoe contact areas on the brake panel. Be sure not to get any grease on the brake drum or linings.*

   b) *Secure the brake shoe retaining washer with new cotter pins.*

11.5 Check the brake cam and the pivot cups in the anchor pin (arrows) for wear and damage

---

12 Rear brake panel - removal, inspection and installation

Removal

350 models

Refer to illustration 12.5

1 Refer to Sections 10 and 11 and remove the brake drum and shoes.

2 Refer to Section 13 and disconnect the brake cables from the brake panel lever.

3 Disconnect the breather hose from the brake panel.

4 Remove and discard the brake panel nuts. Use new nuts on installation.

5 Pull the brake panel away from the axle housing and take it off the axle shaft **(see illustration)**.

250 models

6 The rear brake panel on 250 models is removed as part of the drum removal procedure (see Section 10).

Inspection

Refer to illustrations 12.7, 12.9a, 12.9b, 12.10a, 12.10b, 12.10c and 12.10d

7 Spin the bearing in the brake panel with a finger **(see illustration)**. If it's rough, loose or noisy, replace it as described below. Also check the O-ring and seal for wear or damage and replace them if there's any doubt about their condition.

12.5 Pull off the brake panel; replace the O-ring) (arrow) if there's any doubt about its condition

12.7 If the bearing in the center of the brake panel is rough, loose or noisy, replace it; pry out the seal if it shows signs of leakage

12.9a Remove the snap-ring (arrow) . . .

12.9b . . . . and tap the bearing out with a bearing driver or socket; tap the new bearing in with a bearing driver or socket that bears against the bearing outer race

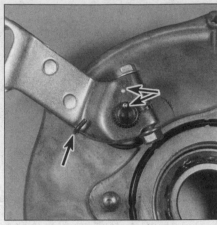

12.10a Look for alignment marks on the cam and lever (arrows); make your own marks if necessary and unhook the spring (arrow)

8   Pry out the seal with a removal tool or screwdriver **(see illustration 12.7)**.

9   Remove the snap-ring and tap out the bearing from the brake shoe side with a bearing driver or socket **(see illustrations)**. Tap in the new bearing with a bearing driver or socket that bears against the bearing outer race. Install the bearing from the axle housing side of the panel, with its sealed side facing the snap-ring. Drive the bearing just past the snap-ring groove, then install the snap-ring. Position the new seal with its lip facing the bearing and tap it in with the same tool used to install the bearing.

10   If the brake cam is loose in its bore or if the seals show any signs of leakage, remove the brake cam from the panel **(see illustrations)**. Inspect the cam and its bore for wear and replace worn parts, then reverse the disassembly sequence to reinstall the brake cam.

11   Apply grease to the rubber seal and oil to the felt seal on assembly.

Installation

350 models

Refer to illustration 12.14

12   Install a new O-ring in the brake panel groove **(see illustration 12.5)**.

13   Position the brake panel on the axle housing, making sure the O-ring isn't dislodged. Install new brake panel nuts (don't re-use the old ones) and tighten them to the torque listed in this Chapter's Specifications.

14   If the brake panel drain bolt was removed, install it with a new

12.10b Slide the wear indicator off the cam; the wide splines on cam and indicator (arrow) align with each other

sealing washer **(see illustration)** and tighten it to the torque listed in this Chapter's Specifications.

15   The remainder of installation is the reverse of the removal steps.

250 models

16   The rear brake panel on 250 models is installed as part of the drum installation procedure (see Section 10).

12.10c Lift out the felt seal; soak the new one with oil on installation

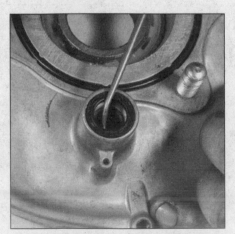

12.10d Pry out the rubber seal from behind the felt seal

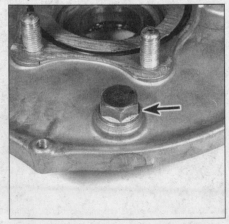

12.14 If the brake panel drain bolt (arrow) is removed, install a new sealing washer

13.1 Remove the wing nuts and spacers, pull the cables rearward and slip them out of the slots (arrows)

13.3 Pull the cable housing back and slip it out of the slot (lower arrow); unhook the pedal spring (upper arrow)

13   Brake pedal, rear brake lever and cables - removal and installation

Brake cables

Removal

Refer to illustration 13.1

1   Unscrew the cable adjusting nuts all the way off the end of the cable, then remove the cables from their slots in the brake panel **(see illustration 10.7 or the accompanying illustration)**.

2   Thread the adjusting hardware and wing nuts back onto the cables so they won't be lost.

Pedal cable

Refer to illustration 13.3

3   Pull the pedal cable housing back from the bracket on the frame near the right swing arm pivot, then slip the cable out through the slot **(see illustration)**.

4   Turn the cable 180-degrees forward and slip it out of the slot in the top of the brake pedal.

Lever cable

5   Pull back the rubber boot from the left handlebar lever.

6   Refer to the brake cable adjustment procedure in Chapter 1 and

back off the adjuster locknut at the handlebar lever and loosen the adjuster all the way.

7   Align the slots in the adjuster and its locknut with each other so they're facing directly away from the handlebar. Turn the cable out of the slots, then align it with the slot in the lever and slip it out.

Installation

8   Installation is the reverse of the removal steps, with the following additions:

a) *Lubricate the cable ends with multi-purpose grease.*
b) *Make sure the cables are secure in their slots and retainers.*
c) *Adjust brake pedal and lever play as described in Chapter 1.*

Brake pedal

Removal

Refer to illustrations 13.11 and 13.12

9   Disconnect the pedal from the cable as described above.

10   Lift the pedal as far as possible and unhook the pedal spring **(see illustration 13.3)**.

11   Straighten the cotter pin, pull it out and remove the washer. Slide the pedal off the shaft **(see illustration)**.

12   Check the seal on each side of the pedal for wear or damage **(see illustration)**. If any defects are visible, pry the seal out and push new ones in.

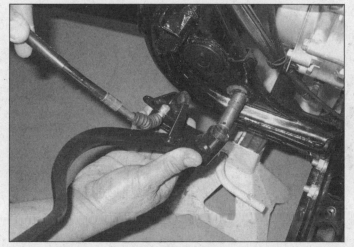

13.11 Slip the pedal off the shaft and disengage the cable end if you haven't already done so

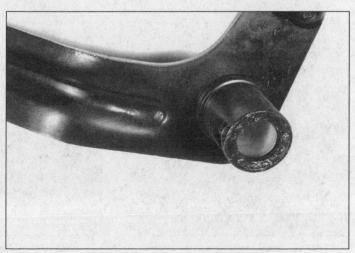

13.12 Inspect the seal on each side of the pedal

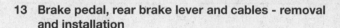

13.16 The punch mark on the handlebar (arrow) aligns with the seam in the lever bracket

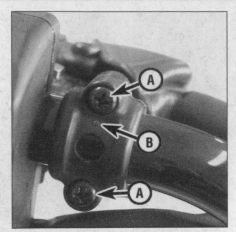

13.17 Remove the clamp screws (A); the dot on the clamp (B) is upward when installed

14.3a Remove the brake pad pin plugs (arrows) . . .

Installation

13   Installation is the reverse of the removal steps, with the following additions:

  a) *Lubricate the pedal shaft and seal lips with multi-purpose grease.*
  b) *Refer to Chapter 1 and adjust brake pedal height.*

Rear brake lever

Removal

Refer to illustrations 13.16 and 13.17

14   Disconnect the cable from the brake lever as described above.

15   Refer to Chapter 2 and disconnect the reverse lock cable.

16   Look for a dot on the handlebar next to the parting line of the lever and clamp **(see illustration)**. This mark is used to position the lever correctly on the handlebar.

17   Remove the lever mounting screws and take it off the handlebar **(see illustration)**.

Installation

18   Installation is the reverse of the removal steps, with the following additions:

  a) *Make sure the handlebar and clamp parting line is aligned with the dot on the handlebar.*
  b) *Install the handlebar clamp with its dot upward (see illustration 13.17). Tighten the upper screw securely, then tighten the lower screw. Don't try to close the gap between the bottom of the clamp and the lever.*

14 Front brake pads (TRX250EX models) - replacement

Refer to illustrations 14.3a, 14.3b, 14.4, 14.5a, 14.5b, 14.5c, 14.6, 14.7a and 14.7b

Warning: *The dust created by the brake system is harmful to your health. Never blow it out with compressed air and don't inhale any of it. An approved filtering mask should be worn when working on the brakes. Do not, under any circumstances, use petroleum-based solvents to clean brake parts. Use brake cleaner only!*

Note: *Always replace all of the front brake pads (on both sides of the vehicle) at the same time.*

1   Refer to Section 18 and remove the front wheel.

2   Disengage the front brake hose from the hose clamp.

3   Unscrew the brake pad pin plugs **(see illustration)**. Loosen the brake pad pins, but don't remove them yet **(see illustration)**. The pins are easier to loosen when the caliper is still bolted to the steering knuckle.

4   Remove the caliper mounting bolts **(see illustration)**.

5   Lift the caliper off, leaving the brake hose connected. **Caution:** *Don't let the caliper hang by the brake hose.* Unscrew the brake pad pins and pull them out, then remove the pads **(see illustrations)**.

6   Remove the pad spring from the caliper **(see illustration)** and inspect it. If it's damaged or distorted, replace it.

7   Remove the shim from the old inner brake pad and install it on the new pad **(see illustrations)**.

8   Installation is the reverse of the removal Steps. Using a c-clamp,

14.3b   and loosen the pins (they're easier to loosen while the caliper is bolted to the steering knuckle)

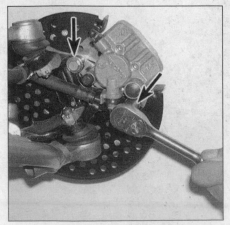

14.4 Unbolt the caliper from the steering knuckle

14.5a Remove the brake pad pins from the caliper

14.5b . . . then remove the outer brake pad . . .

14.5c . . . and the inner brake pad

14.6 Remove the pad spring from the caliper; if it's damaged or distorted, replace it

squeeze the piston back into the caliper bore to make room for the new pads. Tighten the brake pad pins and caliper bolts to the torque listed in this Chapter's Specifications.

9   Replace the pads on the other front caliper.

15 Front brake caliper (TRX250EX models) - removal, overhaul and installation

Warning: *the dust created by the brake system is harmful to your health. Never blow it out with compressed air and don't inhale any of it. An approved filtering mask should be worn when working on the brakes. Do not, under any circumstances, use petroleum-based solvents to clean brake parts. Use brake cleaner only!*

Removal

Refer to illustration 15.1
1   Place a drain pan under the front caliper, then disconnect the brake hose **(see illustration)**. Secure a plastic bag over the end of the hose with a rubber band to prevent fluid loss and keep out dirt. **Note 1:** *If you're removing the caliper to service some other component, leave the brake hose connected. Support the caliper so it doesn't hang by the brake hose.* **Note 2:** *If you're going to overhaul the caliper, but don't have an air compressor to remove the piston (and if the vehicle's hydraulic system is in reasonably good condition), you can remove the piston by pumping the brake lever in Step 5. If you decide to do this, you*

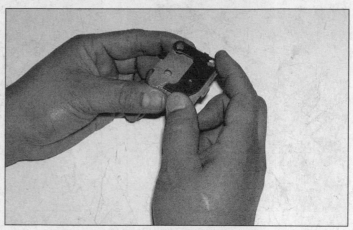

14.7a Remove the shim from the old inner brake pad . . .

leave the brake hose connected for now, then disconnect it after the piston has been removed.
2   Remove the caliper from the steering knuckle. If you're going to overhaul the caliper, remover the brake pads, shim and spring (see Section 14). If you're removing the caliper only to service some other component, leave it assembled and hang it out of the way with a piece of wire. **Caution:** *Support the caliper so it doesn't hang by the brake hose.*

14.7b . . . and install it on the new inner pad

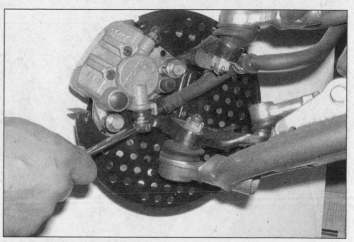

15.1 Remove the brake hose banjo bolt and both sealing washers; discard the washers

15.3a Pry the dust plug from the slide pin hole in the caliper . . .

15.3b . . . unscrew the slide pin from the caliper bracket, remove the bracket . . .

15.3c . . . and remove the slide pin (1), the dust boot (2) and washer (3)

15.4a Remove the dust boot for the pin bolt . . .

15.4b . . . put the caliper bracket in a bench vise, unscrew the pin bolt and remove the washer and indicator plate

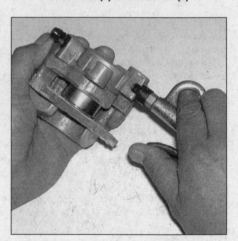

15.6 Slowly apply compressed air to push the piston out of the caliper - DO NOT let your fingers get in the way!

Overhaul

Refer to illustrations 15.3a, 15.3b, 15.3c, 15.4a, 15.4b, 15.6, 15.7, 15.8, 15.9 and 15.10

3   Remove the dust plug from the slide pin hole in the caliper and unscrew the slide pin from the caliper bracket. Remove the caliper bracket and remove the slide pin dust boot and washer **(see illustrations)**.

4   Remove the pin bolt dust boot **(see illustration)**. Place the caliper bracket in a bench vise, unscrew the pin bolt **(see illustration)**, and remove the washer and indicator plate.

5   Remove the pistons from the caliper. If you don't have an air compressor, pump the brake lever to force the pistons out.

6   If you do have a compressor, place some rags between the piston and the caliper frame to act as a cushion. Lay the caliper in the work-

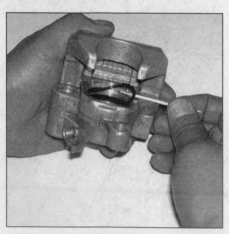

15.7 Using a wood or plastic tool, remove the dust seal

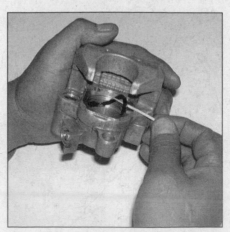

15.8 Using a wood or plastic tool, remove the piston seal

15.9 Remove the bleeder valve and dust cap

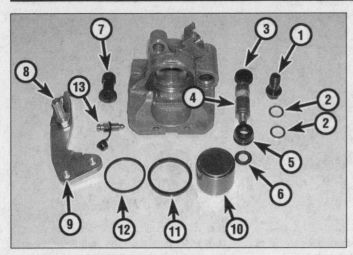

15.10 Front brake caliper details

1	Brake hose banjo bolt	7	Pin bolt dust boot
2	Banjo bolt sealing washer (always replace)	8	Pin bolt
3	Slide pin plug	9	Caliper bracket
4	Slide pin	10	Piston
5	Slide pin dust boot	11	Piston dust seal
6	Slide pin washer	12	Piston seal
		13	Bleed valve and dust cap

bench so the piston is facing down, toward the bench surface. Use compressed air, directed into the fluid inlet, to remove the piston **(see illustration)**. **Caution:** *Use only small quick bursts of air to ease the piston out of the bore. If a piston is blown out with too much force, it could be damaged.* **Warning:** *Never place your fingers in front of the piston. Doing so could result in serious injury.* Once the piston protrudes from the caliper, remove it.

7 Remove the old dust seal **(see illustration)**.

8 Using a wood or plastic tool, remove the piston seal **(see illustration)**.

9 Remove the bleed valve and dust cap **(see illustration)**.

10 Clean the caliper parts with denatured alcohol, fresh brake fluid or brake system cleaner. Dry them off with filtered, unlubricated compressed air and inspect them carefully **(see illustration)**. Replace any obviously worn or damaged parts. Replace all rubber parts (dust boots (piston seals, etc.) whenever you disassemble the caliper. Inspect the surfaces of the piston and piston bore for rust, corrosion, nicks, burrs and loss of plating. If you find defects on the surface of the piston or

piston bore, replace the caliper as an assembly. If the caliper is in bad shape, inspect the master cylinder too (see Section 7).

11 Lubricate the new piston seal with clean brake fluid and install it in its groove in the caliper bore. Make sure it's not twisted and is fully and correctly seated.

12 Install the new dust seal. Make sure that the inner lip of the seal is seated in its groove in the piston and the outer circumference of the seal is seated in its groove in the caliper bore.

13 Lubricate the piston with clean brake fluid and install it into its bore in the caliper. Using your thumbs, push the piston all the way in, making sure it doesn't become cocked in the bore.

14 Install the brake pads and related components (see Section 14).

Installation

15 Install the caliper and brake pads (see Section 14).

16 Connect the brake hose to the caliper, using a new sealing washer on each side of the banjo fitting. Tighten the banjo bolt to the torque listed in this Chapter's Specifications.

17 Remove and overhaul the other front brake caliper.

18 Bleed the front brake system (see Section 8).

16 Front brake disc and hub (TRX250EX) - inspection, removal and installation

1 Remove the front wheels (see Section 17).

Inspection

Refer to illustrations 16.4a and 16.4b

2 Visually inspect the surface of the disc for score marks and other damage. Light scratches are normal after use and won't affect operation, but deep grooves and heavy score marks will reduce braking efficiency and accelerate pad wear. If the disc is badly grooved it must be machined or replaced.

3 To check disc runout, mount a dial indicator with the plunger on the indicator touching the surface of the disc about 1/2-inch from the outer edge. Slowly turn the wheel hub and watch the indicator needle, comparing your reading with the disc runout limit listed in this Chapter's Specifications.

4 The disc must not be thinner than the minimum allowable thickness listed in this Chapter's Specifications (from wear or from machining). If the minimum thickness stamped into the disc differs from the value listed in this Chapter's Specifications, use the specification on the disc **(see illustration)**. Check the thickness of the disc with a micrometer **(see illustration)**. If the disc is thinner than the minimum, replace it.

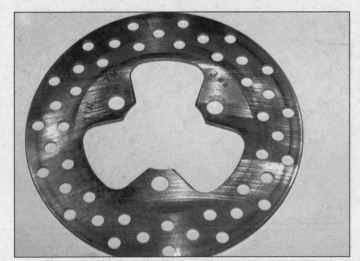

16.4a The minimum thickness of each front disc is stamped into the disc (disc removed for clarity)

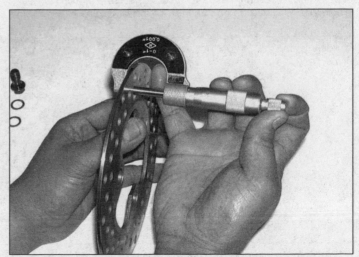

16.4b Check the thickness of the disc with a micrometer (disc removed for clarity)

16.6a Bend back the cotter pin and pull it out of the hub nut

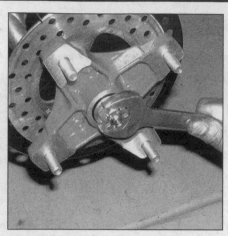

16.6b While an assistant applies the front brake, loosen the hub nut

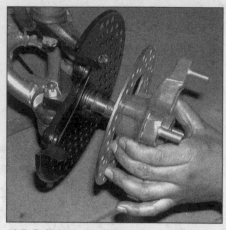

16.7 Pull the hub off the spindle (if the hub is stuck, remove it with a puller)

Removal and installation

Refer to illustrations 16.6a, 16.6b and 16.7

5   Remove the front wheel and brake caliper (see Sections 18 and 15).

6   Bend back the cotter pin and pull it out of the hub nut **(see illustration)**. Have an assistant hold the front brakes on to lock the hub and unscrew the nut. Pull the hub and brake disc off the spindle **(see illustration)**. Discard the cotter pin.

7   Remove the three disc retaining bolts and take the disc off the hub **(see illustration)**. Discard the bolts and use new ones on assembly.

8   Installation is the reverse of the removal Steps, with the following additions:

a) *Use new disc retaining bolts and tighten them to the torque listed in this Chapter's Specifications.*

b) *Tighten the hub nut to the torque listed in this Chapter's Specifications, then install a new cotter pin. If necessary, tighten the nut further to align the cotter pin holes (don't loosen the nut to align the holes).*

17 Front wheel bearings (TRX250EX models) - removal, inspection and installation

Refer to illustrations 17.2, 17.3a, 17.3b, 17.3c, 17.4a, 17.4b, 17.5, 17.6 and 17.7

1   Remove the front brake disc and separate the disc from the hub (see Section 16). **Caution:** *Do not immerse the hub in any kind of cleaning solvent. The sealed hub bearings, which cannot be disassembled*

and repacked, could be damaged if solvent enters them.

2   Spin each bearing with fingers and check for rough, loose or noisy movement **(see illustration)**. If the bearing feels rough or dry, replace it.

3   Remove the outer collar and pry out the old seals **(see illustrations)**. Discard the old seals.

4   The easiest way to remove the hub bearings is to use a suitable puller, similar to the one shown in the accompanying illustration, to remove the outer bearing **(see illustration)**. Then drive out the inner bearing with a bearing removal tool or with a brass drift **(see illustration)**. Pullers can be rented from tool rental dealers. Another method is to insert a drift from one side of the hub, push the spacer aside and place the end of the drift against the inner race of the opposite bearing. Tap the bearing out, working around the circumference. Turn the hub over and remove the spacer, then tap out the remaining bearing. **Caution:** *Once a bearing is removed by pulling or tapping against the inner race, it should not be re-used.*

5   Before assembling the hub, inspect all the parts for wear and damage **(see illustration)**.

6   To install the new bearings, drive them into place with a bearing driver or socket **(see illustration)**. After you've installed the first bearing, don't forget to install the spacer before installing the second bearing. The socket must have an outside diameter that's the same, or slightly smaller than, the outer diameter of the bearings. **Caution:** *Do not strike the center race or the sealed area of the bearing. The resulting damage will cause premature bearing failure.*

7   Tap new inner and outer seals into place with a seal driver or block of wood **(see illustration)**. The open side of each seal faces into the hub. Install the collars in the seals.

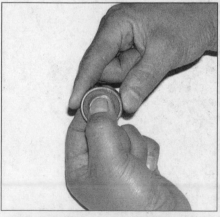

17.2 Spin the bearing with fingers; if it feels rough or dry, replace it (bearing removed from hub for clarity)

17.3a Remove the outer collar . . .

17.3b . . . pry out the outer dust seal . . .

17.3c ... and pry out the inner seal

17.4a A puller like this one is the easiest way to pull out the outer bearing ...

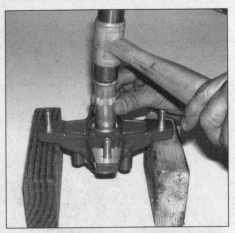

17.4b ... drive out the inner bearing with a bearing removal tool or brass drift

18 Wheels - inspection, removal and installation

Inspection

1   Clean the wheels thoroughly to remove mud and dirt that may interfere with the inspection procedure or mask defects. Make a general check of the wheels and tires as described in Chapter 1.

2   The wheels should be visually inspected for cracks, flat spots on the rim and other damage. Since tubeless tires are involved, look very closely for dents in the area where the tire bead contacts the rim. Dents in this area may prevent complete sealing of the tire against the rim, which leads to deflation of the tire over a period of time.

3   If damage is evident, the wheel will have to be replaced with a new one. Never attempt to repair a damaged wheel.

Removal

4   Securely block the wheel at the opposite end of the vehicle from the wheel being removed, so it can't roll.

5   Loosen the lug nuts on the wheel being removed. Jack up one end of the vehicle and support it securely on jackstands.

6   Remove the lug nuts and pull the wheel off.

Installation

Refer to illustrations 18.7 and 18.8

7   Position the wheel on the studs. Make sure the directional arrow on the tire points in the forward rotating direction of the wheel **(see illustration)**.

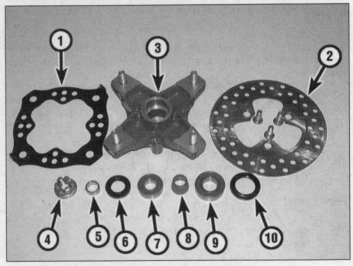

17.5 Front wheel hub details (TRX250EX models)

1	*Splash guard*	*6*	*Outer dust seal*
2	*Front brake disc*	*7*	*Outer bearing*
3	*Front wheel hub*	*8*	*Collar*
4	*Hub nut*	*9*	*Inner bearing*
5	*Collar*	*10*	*Inner dust seal*

17.6 To install the new bearings, drive them into place with a bearing installer tool or an old socket

17.7 Tap new inner and outer seals into place with a block of wood

18.7 The directional arrow on each tire must point in the forward rotating direction of the wheel; if it doesn't, the tire is mounted backward or the wheel is on the wrong side of the vehicle

18.8 Install the wheel nuts with their curved sides toward the wheel

8   Install the wheel nuts with their curved sides toward the wheel **(see illustration)**. This is necessary to locate the wheel accurately on the hub.
9   Snug the wheel nuts evenly in a criss-cross pattern.
10   Remove the jackstands, lower the vehicle and tighten the wheel nuts, again in a criss-cross pattern, to the torque listed in this Chapter's Specifications.

19   Rear wheel hubs - removal and installation

Removal

1   Refer to Section 14 and remove the rear wheel(s).
2   Bend back the cotter pin and pull it out of the hub nut.
3   Unscrew the hub nut and remove the washer.
4   Pull the hub off the axle shaft.

Installation

5   Installation is the reverse of the removal steps, with the following additions:

a) *Lubricate the axle shaft and hub splines with multi-purpose grease.*
b) *Tighten the hub nut to the torque listed in this Chapter's Specifications. If necessary, tighten it an additional amount to align the cotter pin slots.*
c) *Install a new cotter pin and bend it to secure the nut.*

20   Tires - general information

1   Tubeless tires are used as standard equipment on this vehicle. Unlike motorcycle tires, they run at very low air pressures and are completely unsuited for use on pavement. Inflating ATV tires to excessive pressures will rupture them, making replacement of the tire necessary.
2   The force required to break the seal between the rim and the bead

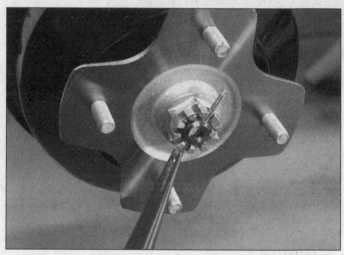

19.2 Bend back the cotter pin and pull it out, then unscrew the nut and remove the hub

of the tire is substantial, much more than required for motorcycle tires, and is beyond the capabilities of an individual working with normal tire irons or even a normal bead breaker. A special bead breaker is required for ATV tires; it produces a great deal of force and concentrates it in a relatively small area.
3   Also, repair of the punctured tire and replacement on the wheel rim requires special tools, skills and experience that the average do-it-yourselfer lacks.
4   For these reasons, if a puncture or flat occurs with an ATV tire, the wheel should be removed from the vehicle and taken to a dealer service department or a repair shop for repair or replacement of the tire. The accompanying illustrations can be used as a guide to tire replacement in an emergency, provided the necessary bead breaker is available.

Turn the tire over and release the other bead.

Deflate the tire and remove the valve core. Release the bead on the side opposite the tire valve with an ATV bead breaker, following the manufacturer's instructions. Make sure you have the correct blades for the tire size (using the wrong size blade may damage the wheel, the tire or the blade). Lubricate the bead with water before removal (don't use soap or any type of lubricant).

If one side of the wheel has a smaller flange, remove and install the tire from that side. Use two tire levers to work the bead over the edge of the rim.

Before installing, ensure that tire is suitable for wheel. Take note of any sidewall markings such as direction of rotation arrows, then work the first bead over the rim flange.

Use tire levers to start the second bead over the rim flange.

Hold the bead while you work the last section of it over the rim flange. Install the valve core and inflate the tire, making sure not to overinflate it.

Notes

Chapter 7
Bodywork and frame

Contents

	Section		Section
Footpegs and mudguards - removal and installation	6	General information	1
Frame - general information, inspection and repair	10	Rear cargo rack - removal and installation	8
Front cargo rack and bumper - removal and installation	5	Rear fender - removal and installation	9
Front fender - removal and installation	7	Seat - removal and installation	2
Fuel tank cover - removal and installation	4	Side covers - removal and installation	3

Specifications

Torque specifications

Footpeg bolts/nuts	33 Nm (24 ft-lbs)
Footpeg bracket bolts	33 Nm (24 ft-lbs)
Step bar mounting bolts (Recon)	33 Nm (24 ft-lbs)

1   General information

Refer to illustration 1.3

This Chapter covers the procedures necessary to remove and install the fenders and other body parts. Since many service and repair operations on these vehicles require removal of the fenders and/or other body parts, the procedures are grouped here and referred to from other Chapters.

In the case of damage to the fenders or other body parts, it is usually necessary to remove the broken component and replace it with a new (or used) one. The material that the fenders and other plastic body parts is composed of doesn't lend itself to conventional repair techniques. There are, however, some shops that specialize in plastic welding, so it would be advantageous to check around first before throwing the damaged part away.

A number of components are secured with reusable plastic retainers **(see illustration)**. To remove a retainer of this type, pry up the center button and pull the retainer out. To install, push the retainer into its hole (with the center button raised), then push the center button down to lock the retainer.

Note: *When attempting to remove any body panel, first study the panel closely, noting any fasteners and associated fittings, to be sure of returning everything to its correct place on installation. In most cases, the aid of an assistant will be required when removing panels, to help*

1.3 Pry up the center post to unlock the retainer (left); push the center post down to lock it in place (right)

avoid damaging the paint. Once the visible fasteners have been removed, try to lift off the panel as described but DO NOT FORCE the panel - if it will not release, check that all fasteners have been removed and try again. Where a panel engages another by means of lugs and grommets, be careful not to break the lugs or to damage the bodywork. Remember that a few moments of patience at this stage will save you a lot of money in replacing broken panels!

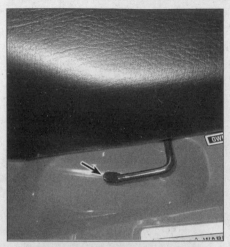

2.1 Lift the seat latch (arrow) and disengage the seat hooks from the bracket

3.2 Free the carburetor vent hose (arrow) from the side cover clips

3.3 Disengage the tabs, then pull the post out of its grommet (arrow)

2   Seat - removal and installation

Refer to illustration 2.1

1   Lift the seat latch lever **(see illustration)** and lift the back end of the seat.
2   Disengage the hooks at the front end of the seat and lift the seat off the vehicle.
3   Installation is the reverse of removal.

3   Side covers - removal and installation

Rancher models

1   Remove the seat (see Section 1).

Right side cover

Refer to illustrations 3.2, 3.3 and 3.4
2   Working inside the side cover, free the carburetor vent hose from the side cover clips **(see illustration)**.
3   Disengage the tabs at the rear and bottom of the recoil starter cover from the fender, then pull the post at the top of the cover out of

its grommet in the lower edge of the side cover **(see illustration)**.
4   Remove the six retainers or screws (see Section 1) that secure the side cover to the fenders and fuel tank cover **(see illustration)**. Pull the side cover up and to the rear to disengage its tabs, then remove it from the vehicle.
5   Installation is the reverse of the removal Steps.

Left side cover

Refer to illustrations 3.6, 3.7a and 3.7b
6   Take the cover off the utility box, then remove the two screws that secure the left side cover to the utility box **(see illustration)**.
7   Remove four trim clips and one bolt that secure the side cover to the fenders and fuel tank cover **(see illustration)**. Pull the side cover up and to the rear to disengage its tabs, then remove it from the vehicle.
8   Installation is the reverse of the removal Steps.

Recon models

9   On 1997 through 2004 models, remove five trim clips (see Section 1) and one bolt. On 2005 models, remove four trim clips.
10   On 1997 through 2004 models, pull the side cover down and back to disengage its tab from the front fender, then remove it from the vehicle.
11   Installation is the reverse of the removal Steps.

3.4 Remove the trim clips (arrows), unscrew the bolt and pull the side cover up and to the rear

3.6 Remove the two screws that secure the left side cover to the utility box (arrows)

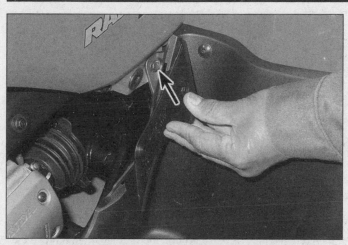

3.7a Remove the cover to expose this trim clip (arrow) . . .

3.7b . . . remove the trim clips and bolt (arrows), then pull the side cover up and to the rear

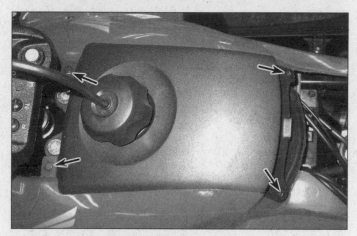

4.2 Remove the trim clip from each corner (arrows), then slide the cover rearward to disengage its tabs

TRX250EX models

12 Remove the four trim clips (see Section 1) that attach the side cover to the front and rear fenders. If the fuel tank cover has not been removed, remove the trim clip that secures the side cover to the fuel tank cover.

13 Pull the side cover down and back to disengage its tab from the front fender, then remove it from the vehicle.

14 Installation is the reverse of the removal Steps.

4 Fuel tank cover - removal and installation

Refer to illustration 4.2

1 Remove the seat and side covers (see Sections 2 and 3). Pull the fuel tank breather hose out of its hole in the handlebar cover, then unscrew the fuel filler cap.

2 Remove the trim clips (see Section 1) from each of the four corners of the cover **(see illustration)**. Slide the cover rearward to disengage its tabs and lift it off the vehicle.

3 Installation is the reverse of the removal Steps.

5 Front cargo rack and bumper - removal and installation

Rancher and Recon models

Refer to illustrations 5.1, 5.2a and 5.2b

1 Working beneath the fender, follow the headlight wiring harness to the tie-wrap that secures it to the bumper and remove the tie wrap **(see illustration)**.

2 **Note:** *This step will be easier, and you'll be less likely to scratch the fender, if you have an assistant.* Remove the cargo rack bolts **(see illustration)**, then lift the cargo rack off the vehicle. Remove the bumper bolts on each side of the vehicle, then lift the bumper off **(see illustration)**.

3 Installation is the reverse of the removal Steps.

5.1 Remove the tie wrap (arrow) that secures the headlight wiring harness to the bumper

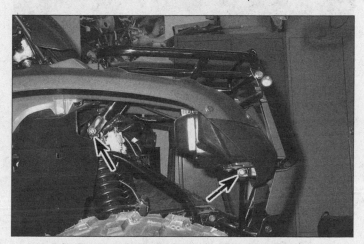

5.2a Remove the bolts (arrows), then lift the cargo rack off the vehicle

5.2b Remove the bolts (arrows), then lift the bumper off the vehicle

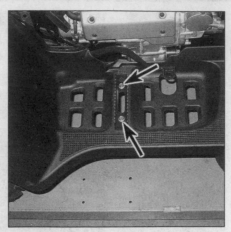

6.2 Unbolt the footpeg, then remove the mudguard trim clips and bolt

6.5 Unbolt the footpeg, then remove the mudguard trim clips and bolt

TRX250EX models

4   These models have a front bumper but are not equipped with a cargo rack.

5   To remove the bumper, unscrew its four bolts and take it off the vehicle.

6   Installation is the reverse of the removal Steps.

---

6   Footpegs and mudguards - removal and installation

Rancher models

Right center mudguard and footpeg

Refer to illustration 6.2

1   Remove the recoil starter cover (see Section 3).

2   Unbolt the footpeg and take it off the vehicle **(see illustration)**.

3   Remove six trim clips (see Section 1) and one bolt, then take the mudguard off the vehicle.

4   Installation is the reverse of the removal Steps.

Left center mudguard and footpeg

Refer to illustration 6.5

5   Unbolt the footpeg and take it off the vehicle **(see illustration)**.

6   Remove five trim clips (see Section 1) and one bolt, then take the mudguard off the vehicle.

7   Installation is the reverse of the removal Steps.

Front mudguards

Refer to illustrations 6.9a and 6.9b

8   Remove the center mudguard as described above.

9   Remove five trim clips (see Section 1) and take the mudguard off the vehicle **(see illustrations)**.

10   Installation is the reverse of the removal Steps.

Recon models

Front mudguards

11   On 1997 through 2004 models, remove four retaining clips (see Section 1) and one screw. On 2005 models, remove five trim clips. Take the mudguard off the vehicle.

12   Installation is the reverse of the removal Steps.

Rear mudguards (1997 through 2004 models)

13   Remove the two screws. Remove the retaining clips (see Section 1). There are four of these if the side cover has been removed and five if it hasn't. Take the mudguard off the vehicle.

14   Installation is the reverse of the removal Steps.

Center mudguards and footpegs

1997 through 2004 models

15   If you're working on the right side of the vehicle, remove the two retaining clips (see Section 1) and detach the engine sub-guard from the vehicle.

16   Remove six retaining clips (see Section 1) and one bolt.

6.9a Remove the trim clips and take the mudguard (one clip hidden) . . .

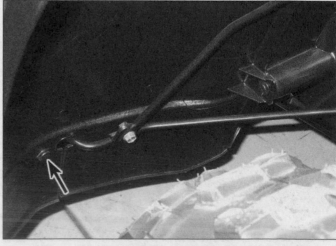

6.9b . . . the hidden clip is accessible from underneath

7.6 Remove the trim clips and unscrew two bolts from each side (arrows)

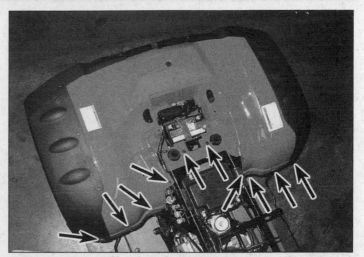

9.4 Remove the bolts and trim clips (arrows) and lift the fender off

17   Unbolt the footpeg, then take the footpeg and mudguard off the vehicle.
18   Installation is the reverse of the removal Steps.

2005 models
19   Remove the front mudguard as described above.
20   Remove four trim clips (see Section 1) and two bolts, then lift the mudguard off.
21   Installation is the reverse of the removal Steps.

TRX250EX models
22   Remove one trim clip (see Section 1) and four screws. Take the mudguard off the vehicle.
23   Unscrew the bolt from the underside of the footpeg. Unbolt the footpeg and take it off the vehicle.
24   If necessary, unbolt the mudguard stays and take them off the vehicle.
25   Installation is the reverse of the removal Steps.

---

7   Front fender - removal and installation

---

1   The front fender is a one-piece unit that spans the front of the vehicle and covers both front tires. A separate flap is attached to each side of the center unit.

Rancher models
Refer to illustration 7.6
2   Remove seat and the side and fuel tank covers (see Sections 2, 3 and 4).
3   Remove the mudguards (see Section 6).
4   Remove the front cargo rack (see Section 5).
5   Disconnect the electrical connectors for the headlight and electrical accessory socket. Free the wiring harnesses from their retainers.
6   Remove six trim clips (see Section 1) and unscrew four bolts **(see illustration)**. With the aid of an assistant, lift the fender off the vehicle. It may be necessary to pull the fender flaps apart slightly to clear the vehicle, but don't separate them any more than absolutely necessary.
7   Installation is the reverse of removal.

Recon models
8   Remove seat and the side and fuel tank covers (see Sections 2, 3 and 4).
9   Remove the front bumper and cargo rack (see Section 5).
10   Disconnect the electrical connectors for the headlights.
11   If you're working on a 1997 through 2004 model, remove two bolts. If you're working on a 2005 model, remove four trim clips, four

bolts and the fender stays. With the aid of an assistant, lift the fender off the vehicle. It may be necessary to pull the fender flaps apart slightly to clear the vehicle, but don't separate them any more than absolutely necessary.
12   Installation is the reverse of removal.

TRX250EX models
13   Remove seat and the side and fuel tank covers (see Sections 2, 3 and 4).
14   Disconnect the electrical connector for the headlights and free the harness from its retainer.
15   Working beneath the fender, remove the four fender bolts. With the aid of an assistant, lift the fender off the vehicle. It may be necessary to pull the fender flaps apart slightly to clear the vehicle, but don't separate them any more than absolutely necessary.
16   Installation is the reverse of removal.

---

8   Rear cargo rack - removal and installation

---

1   A rear cargo rack is used on Rancher and Recon models.
2   Remove the rear cargo rack's mounting bolts and lift the cargo rack off the vehicle. The two smaller bolts have collars.
3   Installation is the reverse of the removal Steps.

---

9   Rear fender - removal and installation

Rancher models
Refer to illustration 9.4
1   Remove the seat and rear cargo rack (see Sections 2 and 8).
2   Remove the battery and starter relay (see Chapter 8). Free the main wiring harness from its retainer and remove the fuse block from the battery box.
3   Follow the taillight wiring harness to its connector and unplug it.
4   Remove ten trim clips (see Section 1) and two bolts that secure the rear fender **(see illustration)**. With the aid of an assistant, lift the fender off, at the same time pulling the starter and battery cables out through the openings in the fender.

Recon models
5   Remove the rear mudguards and cargo rack (see Sections 6 and 8).
6   Lift off the toolbox cover.
7   Remove the battery (see Chapter 8). Disconnect the electrical

connectors for the taillights, main fuse and starter relay (see Chapter 8 if necessary). Make sure the main harness is free of any retainers, then remove it from the fender.

8   Remove the two retaining clips (see Section 1). With an assistant supporting one side of the rear fender unit, lift it off the vehicle, taking care not to scratch the plastic.

9   Installation is the reverse of the removal Steps.

TRX250EX models

10   Remove the seat and side covers (see Sections 2 and 3).

11   Remove the battery (see Chapter 8).

12   Follow the wiring harness from the taillight and disconnect it. Disconnect the wires from the starter relay (see Chapter 8 if necessary). Remove the wiring from the rear fender, together with the diode and fuse block.

13   Remove the four fender bolts. With an assistant supporting one side of the rear fender unit, lift it off the vehicle, taking care not to scratch the plastic.

14   Installation is the reverse of the removal Steps.

10 Frame - general information, inspection and repair

1   All models use a double-cradle frame made of cylindrical steel tubing.

2   The frame shouldn't require attention unless accident damage has occurred. In most cases, frame replacement is the only satisfactory remedy for such damage. A few frame specialists have the jigs and other equipment necessary for straightening the frame to the required standard of accuracy, but even then there is no simple way of assessing to what extent the frame may have been overstressed.

3   After the machine has accumulated a lot of miles, the frame should be examined closely for signs of cracking or splitting at the welded joints. Corrosion can also cause weakness at these joints. Loose engine mount bolts can cause ovaling or fracturing to the engine mounting points. Minor damage can often be repaired by welding, depending on the nature and extent of the damage.

4   Remember that a frame that is out of alignment will cause handling problems. If misalignment is suspected as the result of an accident, it will be necessary to strip the machine completely so the frame can be thoroughly checked.

Chapter 8
Electrical system

Contents

Section

Alternator stator coils, rotor and rear crankcase
   cover - check and replacement 24
Battery - charging.. 4
Battery - check...See Chapter 1
Battery - inspection and maintenance 3
Brake light switches - check and replacement 10
Charging system - output test.. 23
Charging system testing - general information and precautions .. 22
Electric shift system (TE/FE models) - check
   and component replacement.. 26
Electrical troubleshooting... 2
Fuses - check and replacement 5
General information... 1
Handlebar switches - check... 14
Handlebar switches - removal and installation 15
Headlight - bulb replacement... 7
Headlight aim - check and adjustment 8

Section

Ignition main (key) switch - check and replacement 13
Indicator bulbs - replacement 11
Instrument cluster - check, removal and installation.................... 27
Lighting system - check 6
Neutral and reverse switches - check and replacement.............. 16
Oil cooling system - check and component replacement............ 12
Recoil starter ....See Chapter 2
Regulator/rectifier - check and replacement............................ 25
Starter clutch and reduction gears - removal,
   inspection and installation 21
Starter motor - disassembly, inspection and reassembly 20
Starter motor - removal and installation.. 19
Starter relay - check and replacement 18
Starter switch - check and replacement 17
Taillight and brake light bulbs - replacement 9
Wiring diagrams 28

Specifications

Battery

Type	Maintenance free
Capacity	
Rancher..	12V, 12Ah
Recon..	12V-10Ah
TRX250EX...	12V-8 Ah
Current leakage limit	
Rancher	
Without digital instruments.................	0.1 mA
With digital instruments......................	1.0 mA
Recon, TRX250EX..	1.0 mA
Terminal voltage (fully charged)	13.0 to 13.2 volts

Bulbs

Headlights	
Rancher	30/30 watts
Recon	25/25 watts
TRX250EX	35/35 watts
Taillight(s)	5 watts
Brake/taillight(s) (if equipped)........................	5/21 watts
Brake light (separate type, if equipped)	21cp
Indicator lights (bulb type)...............................	1.7 watts

Charging system

Charging output voltage	Less than 15.5 volts at 5000 rpm
Stator coil resistance	0.1 to 1.0 ohms at 20-degrees C (68-degrees F)

Starter motor

Brush length
    All except 2008 and later Recon

Standard	12.5 mm (0.49 inch)
Minimum	9.0 mm (0.35 inch)

    2008 and later Recon

Standard	12.0 mm (0.47 inch)
Minimum	6.5 mm (0.26 inch)
Commutator diameter	Not specified

Fuse ratings

Rancher

Main fuse	30 amps
Lighting fuse	15 amps
Cooling fan and horn	15 amps
Ignition	10 amps
Accessory	10 amps
Electric shift motor (if equipped)	30 amps

Recon

Main fuse	15 amps
Electric shift system (if equipped)	30 amps

TRX250EX

Main fuse	15 amps

Electric shift system

Angle sensor neutral resistance	4 to 6 ohms

Torque specifications

Alternator rotor bolt

Rancher models	108 Nm (80 ft-lbs)
Recon and TRX250EX models	74 Nm (54 ft-lbs)

Stator coil bolts

Rancher models	10 Nm (84 inch-lbs)
Recon and TRX250EDX models	Not specified
Pulse generator bolts	6 Nm (52 inch-lbs)
Reverse/neutral switch bolt	12 Nm (108 inch-lbs)*

Starter clutch Torx bolts

Rancher models	23 Nm (17 ft-lbs)*
Recon and TRX250EX models	
Oil temperature sensor	18 Nm (13 ft-lbs)

Apply non-permanent thread locking agent to the threads.

1   General information

The machines covered by this manual are equipped with a 12-volt electrical system. The components include a three-phase permanent magnet alternator and a regulator/rectifier unit. The regulator/rectifier unit maintains the charging system output within the specified range to prevent overcharging and converts the AC (alternating current) output of the alternator to DC (direct current) to power the lights and other components and to charge the battery.

An electric starter mounted to the engine case behind the cylinder is standard equipment. A kickstarter is also installed. The starting system includes the motor, the battery, the relay and the various wires and switches. If the engine kill switch and the main key switch are both in the On position, the circuit relay allows the starter motor to operate only if the transmission is in Neutral.

Rancher TE and FE models replace the shift pedal with an electrical shifter, operated by a lever on the left handlebar. **Note:** *Keep in mind that electrical parts, once purchased, can't be returned. To avoid unnecessary expense, make very sure the faulty component has been positively identified before buying a replacement part.*

2   Electrical troubleshooting

A typical electrical circuit consists of an electrical component, the switches, relays, etc. related to that component and the wiring and connectors that hook the component to both the battery and the frame. To aid in locating a problem in any electrical circuit, wiring diagrams are included at the end of this Chapter.

Before tackling any troublesome electrical circuit, first study the appropriate diagrams thoroughly to get a complete picture of what makes up that individual circuit. Trouble spots, for instance, can often be narrowed down by noting if other components related to that circuit are operating properly or not. If several components or circuits fail at one time, chances are the fault lies in the fuse or ground/earth connection, as several circuits often are routed through the same fuse and ground/earth connections.

Electrical problems often stem from simple causes, such as loose or corroded connections or a blown fuse. Prior to any electrical troubleshooting, always visually check the condition of the fuse, wires and connections in the problem circuit.

If testing instruments are going to be utilized, use the diagrams to plan where you will make the necessary connections in order to accurately pinpoint the trouble spot.

The basic tools needed for electrical troubleshooting include a test light or voltmeter, a continuity tester (which includes a bulb, battery and set of test leads) and a jumper wire, preferably with a circuit breaker incorporated, which can be used to bypass electrical components. Specific checks described later in this Chapter may also require an ohmmeter.

Voltage checks should be performed if a circuit is not functioning properly. Connect one lead of a test light or voltmeter to either the negative battery terminal or a known good ground/earth. Connect the other lead to a connector in the circuit being tested, preferably nearest to the battery or fuse. If the bulb lights, voltage is reaching that point, which means the part of the circuit between that connector and the battery is problem-free. Continue checking the remainder of the circuit in the same manner. When you reach a point where no voltage is present, the problem lies between there and the last good test point. Most of the time the problem is due to a loose connection. Since these vehicles are designed for off-road use, the problem may also be water or corrosion in a connector. Keep in mind that some circuits only receive voltage when the ignition key is in the On position.

One method of finding short circuits is to remove the fuse and connect a test light or voltmeter in its place to the fuse terminals. There should be no load in the circuit. Move the wiring harness from side-to-side while watching the test light. If the bulb lights, there is a short to ground/earth somewhere in that area, probably where insulation has rubbed off a wire. The same test can be performed on other components in the circuit, including the switch.

A ground check should be done to see if a component is grounded properly. Disconnect the battery and connect one lead of a self-powered test light (such as a continuity tester) to a known good ground/earth. Connect the other lead to the wire or ground/earth connection being tested. If the bulb lights, the ground/earth is good. If the bulb does not light, the ground/earth is not good.

A continuity check is performed to see if a circuit, section of circuit or individual component is capable of passing electricity through it. Disconnect the battery and connect one lead of a self-powered test light (such as a continuity tester) to one end of the circuit being tested and the other lead to the other end of the circuit. If the bulb lights, there is continuity, which means the circuit is passing electricity through it properly. Switches can be checked in the same way.

Remember that all electrical circuits are designed to conduct electricity from the battery, through the wires, switches, relays, etc. to the electrical component (light bulb, motor, etc.). From there it is directed to the frame (ground/earth) where it is passed back to the battery. Electrical problems are basically an interruption in the flow of electricity from the battery or back to it.

3 Battery - inspection and maintenance

Refer to illustration 3.4

1 Most battery damage is caused by heat, vibration, and/or low electrolyte levels, so keep the battery securely mounted, and make sure the charging system is functioning properly. The battery used on these vehicles is a maintenance free (sealed) type and therefore doesn't require the addition of water. However, the following checks should still be regularly performed. **Warning:** *Always disconnect the negative cable first and connect it last to prevent sparks which could cause the battery to explode.*

2 Refer to Chapter 1 for battery removal procedures.

3 Check the battery terminals and cables for tightness and corrosion. If corrosion is evident, disconnect the cables from the battery, disconnecting the negative (-) terminal first, and clean the terminals and cable ends with a wire brush or knife and emery paper. Reconnect

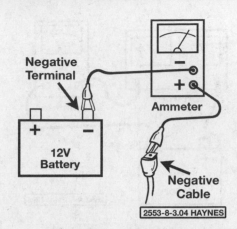

3.4 Checking for a battery drain with an ammeter

the cables, connecting the negative cable last, and apply a thin coat of petroleum jelly to the cables to slow further corrosion.

4 The battery case should be kept clean to prevent current leakage, which can discharge the battery over a period of time (especially when it sits unused). Wash the outside of the case with a solution of baking soda and water. Do not get any baking soda solution in the battery cells. Rinse the battery thoroughly, then dry it. If the battery continually runs down while the vehicle is not being used, there may be a short circuit that's draining the battery. To check, make sure the ignition switch is OFF, then disconnect the negative cable. Connect the positive terminal of an ammeter to the battery negative terminal and connect the ammeter negative terminal to the battery cable **(see illustration). Caution:** *Start with the ammeter at its highest range, then lower the setting until current flow registers. Otherwise, a high current flow may damage the ammeter.* If the current flow exceeds the current drain limit listed in this Chapter's Specifications, disconnect each of the vehicle's circuits in turn until you find the one with the short.

5 Look for cracks in the case and replace the battery if any are found. If acid has been spilled on the frame or battery box, neutralize it with a baking soda and water solution, then touch up any damaged paint. Make sure the battery vent tube (if equipped) is directed away from the frame and is not kinked or pinched.

6 If acid has been spilled on the frame or battery box, neutralize it with the baking soda and water solution, dry it thoroughly, then touch up any damaged paint. Make sure the battery vent tube (if equipped) is directed away from the frame and is not kinked or pinched.

7 If the vehicle sits unused for long periods of time, disconnect the cables from the battery terminals. Refer to Section 4 and charge the battery approximately once every month.

8 The condition of the battery can be assessed by measuring the voltage present at the battery terminals (open circuit voltage); voltmeter positive probe to the battery positive terminal and the negative probe to the negative terminal. When fully charged there should be approximately 13 volts present. If the voltage falls below 12.3 volts the battery must be removed, disconnecting the negative cable first, and charged as described in Section 4.

9 Refer to Chapter 1 to install the battery.

4 Battery - charging

Refer to illustration 4.2

1 If the machine sits idle for extended periods or if the charging system malfunctions, the battery can be charged from an external source.

2 The battery should be charged at no more than the rate printed on the charging rate and time label fixed to the battery. Honda recommends a special battery tester and charger which are unlikely to be available to the vehicle owner. To measure the charging rate, connect

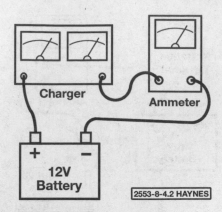

4.2 If the charger doesn't have an ammeter built in, connect one in series as shown; DO NOT connect the ammeter between the battery terminals or it will be ruined

5.1 The fuse box (if equipped) is located near the battery (arrow)

an ammeter in series with a battery charger **(see illustration)**.

3   When charging the battery, always remove it from the machine.

4   Disconnect the battery cables (negative cable first), then connect a voltmeter between the battery terminals and measure the voltage.

5   If terminal voltage is within the range listed in this Chapter's Specifications, the battery is fully charged. If it's lower, recharge the battery.

6   A quick charge can be used in an emergency, provided the maximum charge rate and time printed on the battery are not exceeded (exceeding the maximum rate or time may buckle the battery plates, rendering it useless). A quick charge should always be followed as soon as possible by a charge at the standard rate and time.

7   Hook up the battery charger leads (positive lead to battery positive terminal, negative lead to battery negative terminal), then, and only then, plug in the battery charger. **Warning:** *The hydrogen gas escaping from a charging battery is explosive, so keep open flames and sparks well away from the area. Also, the electrolyte is extremely corrosive and will damage anything it comes in contact with.*

8   Allow the battery to charge for the specified time. If the battery overheats or gases excessively, the charging rate is too high. Either disconnect the charger or lower the charging rate to prevent damage to the battery.

9   After the specified time, unplug the charger first, then disconnect the leads from the battery.

10   If the recharged battery discharges rapidly when left disconnected, it's likely that an internal short caused by physical damage or sulfation has occurred. A new battery will be required. A sound battery

will tend to lose its charge at about 1% per day.

11   When the battery is fully charged, unplug the charger first, then disconnect the leads from the battery. Wipe off the outside of the battery case and install the battery in the vehicle.

5   Fuses - check and replacement

Refer to illustrations 5.1a, 5.1b, 5.1c and 5.1d

1   All models have a main fuse in the wire from the battery to the ignition switch **(see illustrations)**. Additional fuses used on some models are located near the battery.

2   The fuses can be removed and checked visually. To check, remove the seat (see Chapter 7) and open the cover. A blown fuse is easily identified by a break in the element.

3   If a fuse blows, be sure to check the wiring harnesses very carefully for evidence of a short circuit. Look for bare wires and chafed, melted or burned insulation. If a fuse is replaced before the cause is located, the new fuse will blow immediately.

4   Never, under any circumstances, use a higher rated fuse or bridge the fuse terminals, as damage to the electrical system could result.

5   Occasionally a fuse will blow or cause an open circuit for no obvious reason. Corrosion of the fuse ends and fuse holder terminals may occur and cause poor fuse contact. If this happens, remove the corrosion with a wire brush or emery paper, then spray the fuse end and terminals with electrical contact cleaner.

5.1b Some fuses (arrows) are mounted in individual holders . . .

5.1c . . . secured in a slot by a tab (A); lift the lever (B) . . .

5.1d ... and pull off the cover to expose the fuse (arrow), which can then be pulled out

6   Lighting system - check

1   The battery provides power for operation of the headlights, taillight, brake light (if equipped) and instrument cluster lights. If none of the lights operate, always check battery voltage before proceeding. Low battery voltage indicates either a faulty battery, low battery electrolyte level or a defective charging system. Refer to Chapter 1 and Section 3 of this Chapter for battery checks and Sections 22 and 23 for charging system tests. Also, check the condition of the fuses and replace any blown fuses with new ones.

Headlights

2   If both of the headlight bulbs are out with the lighting switch in the On position, check the light fuse (models so equipped) (see Section 5).
3   If only one headlight is out, refer to Section 7 and disconnect the electrical connector for the headlight bulb. Use a jumper wire to connect the bulb directly to the battery terminals as follows:

a) *Green wire (ground) terminal to battery negative terminal*
b) *White wire (low beam) terminal to battery positive terminal*
c) *Blue/black wire (high beam) terminal to battery positive terminal.*

   If the light comes on, the problem lies in the wiring or one of the switches in the circuit. Refer to Sections 14 and 15 for the switch testing procedures, and also the wiring diagrams at the end of this manual. If either filament (high or low beam) of the bulb doesn't light, the bulb is burned out. Refer to Section 7 and replace it.

Taillight

4   If the taillight fails to work, check the bulb and the bulb terminals first, then check for battery voltage at the power wire in the taillight. If voltage is present, check the ground circuit for an open or poor connection.
5   If no voltage is indicated, check the wiring between the taillight and the lighting switch, then check the switch.

Brake light

6   See Section 10 for the brake light circuit checking procedure.

Neutral or reverse indicator light

Note: *This section applies to models with indicator bulbs. Non-replaceable LED indicators are used with digital instrument clusters.*
7   If the light fails to operate when the transmission is in Neutral or Reverse, check the light fuse (350 models only) and the bulb (see Section 11 for bulb removal procedures). If the bulb and fuse are in good condition, check for battery voltage at the wire attached to the switch (see Section 16 for the switch location). If battery voltage is present, refer to Section 16 for the switch check and replacement procedures.
8   If no voltage is indicated, check the wiring to the bulb, to the switch and between the switch and the bulb for open circuits and poor connections.

7   Headlight - bulb replacement

Warning: *If the headlight has just burned out, give it time to cool before changing the bulb to avoid burning your fingers.*
Refer to illustrations 7.1, 7.2a and 7.2b
1   On Rancher and Recon models, reach inside the front fender, remove the headlight cover screw (if equipped) and take off the cover **(see illustration)**.
2   If you're working on a 2000 through 2003 2WD Rancher or any Recon, pull the rubber dust cover off the headlight case **(see illustration)**. Push the bulb socket forward and turn it counterclockwise to release it, then pull it out **(see illustration)**. Pull the bulb out of the socket, noting the position of the alignment tab on the base of the bulb. Install the new bulb, aligning the tab, then reverse the removal steps to complete the installation. The Top mark on the dust cover goes upward.
3   If you're working on a 2000 through 2003 4WD Rancher, twist the bulb socket counterclockwise and pull it out of the housing. Unlatch the connector and disconnect it from the bulb. Check the bulb O-ring and replace it if necessary. Install by reversing the removal steps.
4   If you're working on a 2004 or later Rancher, pull the dust cover off the headlight housing. Disconnect the electrical connector from the bulb, turn the bulb counterclockwise and pull it out of the headlight

7.1 Remove the headlight cover (arrow)

7.2a Pull the tab on the rubber cover (arrow) ...

7.2b ... to pull back the rubber cover and expose the bulb socket ...

housing. Install by reversing the removal steps.
5   If you're working on a TRX250EX, disconnect the electrical connector from the back of the bulb. Twist the bulb counterclockwise and
pull it out of the housing. Check the bulb O-ring and replace it if necessary. Install by reversing the removal steps.

8   Headlight aim - check and adjustment

Refer to illustration 8.3
1   An improperly adjusted headlight may cause problems for oncom-
ing traffic or provide poor, unsafe illumination of the terrain ahead.
Before adjusting the headlight, be sure to consult with local traffic laws
and regulations. Honda doesn't provide specifications for headlight
adjustment.
2   The headlight beam can be adjusted horizontally. Before performing the adjustment, make sure the fuel tank is at least half full, and
have an assistant sit on the seat.
3   If you're working on a Rancher, turn the adjuster as necessary to
move the beam **(see illustration)**. Turning the bolt clockwise will move
the beam toward the center of the vehicle. Turning it counterclockwise
will move it away from the center.
4   If you're working on a Recon, turn the adjusting screw, which is
located in the bottom center of the headlight housing and accessible
from the front, in or out to move the beam vertically.
5   If you're working on a TRX250EX, loosen the vertical adjusting
screw in the bottom of the headlight housing. Slide the screw back and
forth in the slot to change the adjustment, then tighten the screw.

9   Taillight and brake light bulbs - replacement

Taillight
Refer to illustrations 9.2 and 9.4
1   If you're working on a Rancher, remove the screw from the bulb
cover, then squeeze the cover to free its tabs **(see illustration)**.
2   If you're working on a Recon, open the toolbox for access to the
bulb socket (1997 through 2004). 2005 Recon models use a light-emit-
ting diode (LED) taillight, so the entire assembly (lens and housing)
must be replaced if the LEDs burn out.
3   If you're working on a TRX250EX, tug the taillight housing to free
its posts from the grommets in the rear fender.
4   On all except 2005 Recon models, twist the bulb socket and pull
it out of its housing, then pull the bulb out of the socket **(see illustration)**. On 2005 Recon models, work the lamp housing posts free of the
grommets.

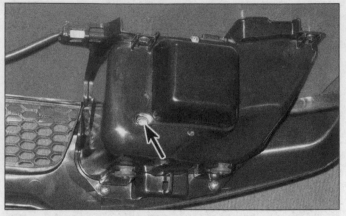

**8.3 On Rancher models, turn the adjuster (arrow)
to move the beam**

5   Check the socket terminals for corrosion and clean them if necessary.
6   Make sure the rubber gasket is in place and in good condition,
then push the bulb into the socket.

Brake light
7   If the vehicle is equipped with a separate brake light, remove the
lens securing screws and take off the lens.
8   Press the bulb into its socket and turn it counterclockwise to
remove.
9   Installation is the reverse of removal.

10   Brake light switches - check and replacement

1   Brake lights are used on some models, as required by local area
regulations.
2   Before checking any electrical circuit, check the fuses (see Section 5).
3   Using a test light connected to a good ground, check for voltage
to the pink wire at the brake light switch. If there's no voltage present,
check the pink wire between the switch and the ignition switch (see the
wiring diagrams at the end of the book).
4   If voltage is available, touch the probe of the test light to the other
terminal of the switch, then pull the brake lever or depress the brake
pedal - if the test light doesn't light up, replace the switch.
5   If the test light does light, check the wiring between the switch
and the brake lights (see the wiring diagrams at the end of the book).

**9.2 Remove the screw, squeeze the cover
and pull it off**

**9.4 Twist the socket and pull it out, then
pull the bulb out**

**10.10 Disconnect the spring from
the switch**

11.1 Pull the bulb socket out of the handlebar cover and pull the lens out of the socket . . .

11.2 . . . then pull the bulb out, using a piece of rubber hose if necessary; align the socket lugs (if equipped) with the cover grooves (arrow) on installation

12.3 The oil temperature sensor (arrow) is on the back of the engine

Switch replacement

Brake lever switch

6   Disconnect the electrical connectors from the switch.
7   Remove the mounting screw and detach the switch from the brake lever bracket (left side) or front master cylinder (right side).
8   Installation is the reverse of the removal procedure. The brake lever switch isn't adjustable.

Brake pedal switch

Refer to illustration 10.10
9   Disconnect the electrical connector in the switch harness.
10   Disconnect the spring from the brake pedal switch **(see illustration)**.
11   Unbolt the switch body and remove the switch.
12   Install the switch by reversing the removal procedure. Position the switch so the brake light comes on when you operate the brake pedal.

11   Indicator bulbs - replacement

Refer to illustrations 11.1 and 11.2
1   This Section applies to models with removable indicator bulbs. To replace a bulb, pull the appropriate rubber socket out of the handlebar cover, then pull off the lens **(see illustration)**.
2   Pull the bulb out of the socket **(see illustration)**. **Note:** *To reach the bulbs, which are deep within the sockets, push a piece of rubber or soft plastic tubing down over the bulb, then pull the tubing and bulb out. Use the same tubing to install the new bulb.*
3   If the socket contacts are dirty or corroded, they should be scraped clean and sprayed with electrical contact cleaner before new bulbs are installed.
4   Carefully push the new bulb into position, using the same tubing that was used for removal, then push the socket into the handlebar cover.

12   Oil cooling system - check and component replacement

1   Rancher models are equipped with an oil temperature warning system and cooling fan (the fan is optional on US 2WD foot shift models). The system turns on an indicator light on the handlebars or instrument cluster when the oil overheats. The system consists of an oil temperature sensor mounted on the underside of the engine, a control unit mounted at the front of the vehicle and the warning indicator light. The indicator light should come on for a few seconds when the engine is first started, then turn off. For operation in high temperatures, the electric fan (if equipped) comes on to cool the oil.

Control system check

Refer to illustration 12.3
Note: *You'll need to remove the right inner fender for access to some of the wiring connectors* (see Chapter 7).
2   If the oil temperature warning light fails to operate properly, check the neutral and reverse indicators. If they don't work either, check the wiring for breaks or poor connections. If the oil light is the only one that doesn't work, check the bulb (if equipped) and replace it if it's burned out. The LED in the digital instrument cluster (if equipped) can't be replaced separately. Refer to the following step to check the overall system and Steps later in this procedure to check the instrument cluster.
3   Locate the oil temperature sensor on the back of the engine near the left lower engine mount **(see illustration)**. **Note:** *Check the wire color in the wiring diagrams at the end of the manual to make sure you're not mistaking the oil temperature sensor for the reverse shift switch. They're near each other and look somewhat alike.* Disconnect its wire, then connect the wire to ground (bare metal on the engine) with a jumper wire. When the key is turned to On, the fan should come on and the warning lamp should light, then they should both go off when the key is turned to Off. If the system performs as described, the sensor is probably defective. If not, disconnect the jumper wire, reconnect the sensor wire and perform the following steps.
4   If the fan doesn't start and the oil light doesn't come on, disconnect the electrical connector from the ignition control module. Using an ohmmeter, check continuity between the light blue wire's terminal (ignition control unit) and the blue wire's terminal (oil temperature sensor). If there's no continuity, check the wiring for breaks or poor connections. If there is continuity, the ignition control unit may be at fault. Since this can't be returned once purchased, it's a good idea to have it tested by a dealer service department or other qualified shop.
5   If the fan motor won't stop running and the light won't go out, disconnect the electrical connector from the ignition control module. Using an ohmmeter, check continuity between the blue wire's terminal (oil temperature sensor) and ground (bare metal on the engine). If there is continuity, check the wiring for breaks or poor connections. If there is no continuity, the ignition control unit may be at fault. Since this can't be returned once purchased, it's a good idea to have it tested by a dealer service department or other qualified shop.
6   If the fan motor doesn't run but the light comes on, disconnect the fan's electrical connectors. Connect them directly to a 12-volt battery (the vehicle's battery will work if it's fully charged). The fan should run. If not, replace it. If it does run, reconnects its wiring connectors.
7   If the fan motor is good, disconnect the single white wire at the ignition control module. Connect the harness side of the wire (not the

ICM side) to ground (nearby bare metal) with a jumper wire. Turn the key to On (but don't start the engine). The fan should run. If it doesn't, check the green wire (ground wire) for breaks or bad connections. If the wire is good, the ICM may be defective. Since this can't be returned once purchased, it's a good idea to have it tested by a dealer service department or other qualified shop.

8   If the white and green wires are good and the fan still doesn't run, disconnect the fan motor's power wire connector (blue wire). With the key On, check for voltage between the wire and ground. If there's no voltage, check the wire for breaks or bad connections. If there is voltage, reconnect the blue wire. Disconnect the white wire from the green wire and check the white wire for breaks or bad connections between the terminal and the ICM. If the wire is good, replace the fan.

9   If the fan motor won't stop running, disconnect the fan motor's white wire-to-green wire connector and disconnect the single white wire connector at the ICM. Check for continuity between the white wire and ground (bare metal). If there is continuity, check the white wire for a short to ground (such as worn insulation allowing the wire to touch bare metal). If there's no continuity, the ignition control unit may be at fault. Since this can't be returned once purchased, it's a good idea to have it tested by a dealer service department or other qualified shop.

10   If the fan works but the oil light doesn't come on, disconnect the electrical connector at the ICM and ground the blue-red wire to bare metal with a jumper wire. If the light now comes on with the key On, the ignition control unit may be at fault. Since this can't be returned once purchased, it's a good idea to have it tested by a dealer service department or other qualified shop.

11   If grounding the wire in Step 9 doesn't make the light come on, check the blue-red wire from the ICM to the digital instrument cluster or warning light panel for a break or bad connection. If the wire is good, the cluster or panel may be at fault. Since this can't be returned once purchased, it's a good idea to have it tested by a dealer service department or other qualified shop.

Oil temperature sensor check

12   Disconnect the electrical connector from the sensor **(see illustration 12.3)**. With the engine cold, connect the ohmmeter between the switch terminal and ground. It should indicate the same reading as in Step 4a above. Now warm up the engine to normal operating temperature. The resistance reading should be considerably higher than in Step 4a.

13   If the sensor doesn't operate as described, replace it. If the sensor is good and the fan still doesn't operate, test it as described below.

Oil temperature sensor replacement

14   Drain the engine oil (see Chapter 1).
15   Unscrew the sensor from the engine.

12.22 Unbolt the fan motor and shroud (arrows) and remove it from the vehicle

16   Installation is the reverse of the removal steps. Use an electrically conductive sealer on the switch threads and tighten it to the torque listed in this Chapter's Specifications.

17   Fill the crankcase with the recommended type and amount of oil (see Chapter 1) and check for leaks.

Fan motor check

18   Start by removing the seat (see Chapter 7) and checking the fan fuse in the fuse box (see Section 5).

19   If the fuse is good, follow the wiring harness from the fan to the connector and disconnect it. Connect the fan motor directly to the battery with a pair of jumper wires. If the fan motor doesn't run, replace it.

Fan motor replacement

Refer to illustrations 12.22, 12.23, 12.24a, 12.24b and 12.26

20   Remove the oil cooler and fuel tank (see Chapters 2 and 3).

21   Disconnect the fan breather tube from the fan and free it from the retainer.

22   Unbolt the fan motor and shroud from the bracket and remove it from the left side of the vehicle **(see illustration)**.

23   Remove the screws and take the grille off the fan **(see illustration)**.

24   Unscrew the nut, pull the fan off the motor and remove the collar and cap **(see illustrations)**.

25   Remove the screws that secure the fan motor to the shroud, then lift off the motor and remove the collars and grommets.

26   Installation is the reverse of the removal steps. Route the harness and breather tube upward **(see illustration)**.

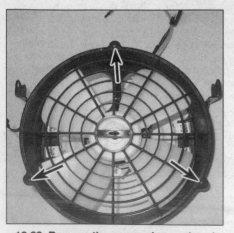

12.23 Remove the screws (arrows) and take the grille off

12.24a Remove the nut (arrow) and fan . . .

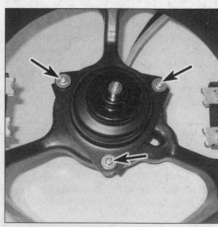

12.24b . . . and the mounting bolts (arrows) to separate the motor from the bracket

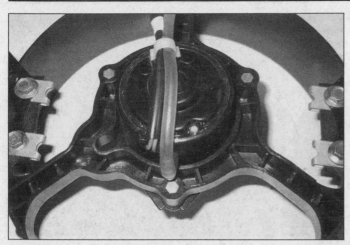

12.26 Route the wiring harness and breather hose like this

13.4 Squeeze the prongs to detach the switch from the handlebar cover

13   Ignition main (key) switch - check and replacement

Check

1   Follow the wiring harness from the ignition switch to the connector and disconnect the connector.

2   Using an ohmmeter, check the continuity of the terminal pairs indicated in the wiring diagrams at the end of the book. Continuity should exist between the terminals connected by a solid line when the switch is in the indicated position.

3   If the switch fails any of the tests, replace it.

Replacement

Refer to illustration 13.4

4   The ignition switch is secured to the handlebar cover by plastic prongs **(see illustration)**.

5   If you haven't already done so, unplug the switch electrical connector. Squeeze the prongs and lift the switch out of the handlebar cover.

6   Installation is the reverse of the removal procedure.

14   Handlebar switches - check

1   Generally speaking, the switches are reliable and trouble-free. Most troubles, when they do occur, are caused by dirty or corroded contacts, but wear and breakage of internal parts is a possibility that should not be overlooked. If breakage does occur, the entire switch and related wiring harness will have to be replaced with a new one, since individual parts are not usually available.

2   The switches can be checked for continuity with an ohmmeter or a continuity test light. Always disconnect the battery negative cable, which will prevent the possibility of a short circuit, before making the checks.

3   Trace the wiring harness of the switch in question and unplug the electrical connectors.

4   Using the ohmmeter or test light, check for continuity between the terminals of the switch harness with the switch in the various positions. Refer to the continuity diagrams contained in the wiring diagrams at the end of the book. Continuity should exist between the terminals connected by a solid line when the switch is in the indicated position.

5   If the continuity check indicates a problem exists, refer to Section 15, disassemble the switch and spray the switch contacts with electrical contact cleaner. If they are accessible, the contacts can be scraped clean with a knife or polished with crocus cloth. If switch components are damaged or broken, it will be obvious when the switch is disassembled.

15   Handlebar switches - removal and installation

Refer to illustration 15.1

1   The handlebar switches are composed of two halves that clamp around the bars. They are easily removed for cleaning or inspection by taking out the clamp screws and pulling the switch halves away from the handlebars **(see illustration)**.

2   To completely remove the switches, the electrical connectors in the wiring harness must be unplugged and the harness separated from the tie wraps and retainers.

3   When installing the switches, make sure the wiring harness is properly routed to avoid pinching or stretching the wires. If there's a locating pin inside the switch, make sure it fits into the hole in the handlebar.

16   Neutral and reverse switches - check and replacement

1   This procedure applies to the external neutral and reverse switches used on 2004 and earlier foot shift Recon models, as well as 2005 and earlier TRX250EX models. The combined gear position switch used on all other models is covered in Section 26.

Check

2   Locate the wiring connector from the reverse and neutral switches under the rear fender (it can be identified by its wire colors).

3   Connect one lead of an ohmmeter to a good ground and the other lead to the terminal for the switch function being tested (light green wire for neutral; gray wire for reverse).

15.1 Remove the screws (arrows) to separate the switch housing halves

16.8a Here's a Rancher reverse switch (arrow) . . .

16.8b . . . and here's the switch on a Recon

4   When the transmission is in neutral, the ohmmeter should read 0 ohms between the neutral switch and ground - in any other gear, the meter should read infinite resistance.

5   When the transmission is in reverse, the ohmmeter should read 0 ohms between the reverse switch and ground - in any other gear, the meter should read infinite resistance.

6   If the switch doesn't check out as described, replace it.

Replacement

Refer to illustrations 16.8a and 16.8b

7   Unbolt the reverse lockout cable cover from the engine (see Chapter 2).

8   Wiggle the wiring connectors and work them off the switches. Unscrew the switch(es) from the engine **(see illustrations)**. **Note:** *Check the wire color in the wiring diagrams at the end of the manual to make sure you're not mistaking the reverse shift switch for the oil temperature sensor. They're near each other and look somewhat alike.*

9   Install the switch in the case, using a new sealing washer. Tighten it to the torque listed in this Chapter's Specifications.

10   Reconnect the wiring connectors, using the R and N labels molded into the grommets to connect them to the correct switches.

11   The remainder of installation is the reverse of the removal steps.

17   Starter switch - check and replacement

The starter is switch is part of the switch assembly on the left handlebar. Refer to Section 14 for checking and replacement procedures.

18   Starter relay - check and replacement

Refer to illustration 18.3

1   Refer to Chapter 7 and remove the battery seat.

2   Disconnect the negative cable from the battery.

3   Pull back the rubber covers from the terminal nuts, remove the nuts and disconnect the starter relay cables **(see illustration)**. Disconnect the remaining electrical connector (thin wires) from the starter relay.

4   Connect an ohmmeter between the terminals from which the nuts were removed. It should indicate infinite resistance.

5   Connect a 12-volt battery to the terminals of the thin wires. The vehicle's battery can be used if it's fully charged. The ohmmeter should now indicate zero ohms.

6   If the relay doesn't perform as described, pull its rubber mount off the metal bracket and pull the relay out of the mount.

7   Installation is the reverse of removal. Reconnect the negative battery cable after all the other electrical connections are made.

19   Starter motor - removal and installation

Removal

Refer to illustrations 19.3 and 19.6

1   Disconnect the cable from the negative terminal of the battery.

2   Remove the right side engine cover.

3   Pull back the rubber boot and remove the nut retaining the starter cable to the starter **(see illustration)**.

18.3 Remove the terminal nuts (arrows) and disconnect the cables, then disconnect the thin wires

19.3 Disconnect the cable ad remove the mounting bolts (arrows), then lift the starter and pull it out (350 shown; 250 similar)

19.6 Pull the starter out and inspect the O-ring (arrow)

20.2 Make alignment marks between the housing and end covers before disassembly

4   Remove the starter mounting bolts.
5   Lift the outer end of the starter up a little bit and slide the starter out of the engine case. **Caution:** *Don't drop or strike the starter or its magnets may be demagnetized, which will ruin it.*
6   Check the condition of the O-ring on the end of the starter and replace it if necessary **(see illustration)**.

Installation

7   Remove any corrosion or dirt from the mounting lugs on the starter and the mounting points on the crankcase.
8   Apply a little engine oil to the O-ring and install the starter by reversing the removal procedure.

---

20 Starter motor - disassembly, inspection and reassembly

1   Remove the starter motor (see Section 19).

Disassembly

Refer to illustrations 20.2 and 20.3

Note: *This procedure shows the starter design used on all except 2008 and later Recon models. The later design is similar, but the brushes contact the end of the commutator, rather than the sides.*

2   Make alignment marks between the housing and covers **(see illustration)**.
3   Unscrew the two long bolts, then remove the cover with its O-ring from the motor. Remove the shim(s) from the armature, noting

their correct locations.
4   Remove the front cover with its O-ring from the motor. Remove the toothed washer from the cover and slide off the insulating washer and shim(s) from the front end of the armature, noting their locations.
5   Withdraw the armature from the housing.

Inspection

Note: *Check carefully which components are available as replacements before starting overhaul.*
Refer to illustrations 20.7, 20.8, 20.9, 20.11a and 20.11b
6   Connect an ohmmeter between the terminal bolt and the insulated brush holder (or the indigo colored wire). There should be continuity (little or no resistance). When the ohmmeter is connected between the rear cover and the insulated brush holder (or the indigo colored wire), there should be no continuity (infinite resistance). Replace the brush holder if it doesn't test as described.
7   Unscrew the nut from the terminal bolt and remove the plain washer, insulating washers and rubber ring, noting carefully how they're installed **(see illustration)**. Withdraw the terminal bolt and brush assembly from the housing and recover the insulator.
8   Lift the brush springs and slide the brushes out of their holders **(see illustration)**.
9   The parts of the starter motor that most likely will require attention are the brushes. If one brush must be replaced, replace both of them. The brushes are replaced together with the terminal bolt and the brush plate. Brushes must be replaced if they are worn excessively, cracked, chipped, or otherwise damaged. Measure the length of the brushes and compare the results to the brush length listed in this Chapter's

20.7 Unscrew the nut and remove the washers from the terminal bolt, noting their correct installed order

20.8 Lift the brush springs and slide the brushes out of their holders

20.9 Measure the brush length and replace the brushes if they're worn

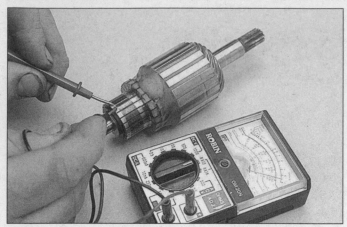

20.11a Check for continuity between the commutator bars . . .

Specifications **(see illustration)**. If either of the brushes is worn beyond the specified limits, replace them both.

10   Inspect the commutator for scoring, scratches and discoloration. The commutator can be cleaned and polished with fine emery paper, but do not use sandpaper and do not remove copper from the commutator. After cleaning, clean out the grooves and wipe away any residue with a cloth soaked in an electrical system cleaner or denatured alcohol.

11   Using an ohmmeter or a continuity test light, check for continuity between the commutator bars **(see illustration)**. Continuity should exist between each bar and all of the others. Also, check for continuity between the commutator bars and the armature shaft **(see illustration)**. There should be no continuity between the commutator and the shaft. If the checks indicate otherwise, the armature is defective.

12   Check the dust seal in the front cover for wear or damage. Check the needle roller bearing in the front cover for roughness, looseness or loss of lubricant. Check with a motorcycle shop or Honda dealer to see if the bearing can be replaced separately; if this isn't possible, replace the starter motor.

13   Inspect the bushing in the rear cover. Replace the starter motor if the bushing is worn or damaged.

14   Check the starter pinion for worn, chipped or broken teeth. If the gear is damaged or worn, replace the starter motor. Inspect the insulating washers for signs of damage and replace if necessary.

Reassembly

Refer to illustrations 20.16, 20.17, 20.19a, 20.19b, 20.20, 20.21, 20.22 and 20.23

15   Lift the brush springs and slide the brushes back into position in their holders.

16   Fit the insulator to the housing and install the brush plate. Insert

20.11b . . . and for no continuity between each commutator bar and the armature shaft

the terminal bolt through the brush plate and housing **(see illustration)**.

17   Slide the rubber ring and small insulating washer onto the bolt, followed by the large insulating washers and the plain washer. Fit the nut to the terminal bolt and tighten it securely **(see illustration)**.

18   Locate the brush assembly in the housing, making sure its tab is correctly located in the housing slot.

19   Fit the shims to the armature shaft **(see illustration)** and insert the armature in the housing, locating the brushes to the commutator bars. Check that each brush is securely pressed against the commutator by its spring and is free to move easily in its holder **(see illustration)**.

20   Fit the lockwasher to the front cover so that its teeth are correctly located with the cover ribs **(see illustration)**. Apply a smear of grease

20.16 Install the brush plate assembly and terminal bolt

20.17 Install the terminal bolt nut and tighten it securely

20.19a Install the shims on the armature, then insert the armature . . .

20.19b . . . and locate the brushes on the commutator

20.20 Fit the lockwasher to the front cover so its teeth engage the cover ribs

20.21 Install the shims and washer on the armature, making sure they're in the correct order

20.22 Fit the O-ring . . .

to the cover dust seal lip.

21   Slide the shim(s) onto the front end of the armature shaft, then fit the insulating washer **(see illustration)**. Fit the sealing ring to the housing and carefully slide the front cover into position, aligning the marks made on removal.

22   Ensure the brush plate inner tab is correctly located in the housing slot and fit the O-ring to the housing **(see illustration)**.

23   Align the rear cover groove with the brush plate outer tab and install the cover **(see illustration)**.

24   Check that the marks made on disassembly are correctly aligned, then fit the long bolts and tighten them securely **(see illustration 20.3)**.

25   Install the starter as described in Section 19.

20.23 . . . and install the rear cover, aligning its groove with the brush plate outer tab

21 Starter clutch and reduction gears - removal, inspection and installation

Starter clutch

Refer to illustrations 21.2, 21.3, 21.4a and 21.4b

1   Remove the alternator cover and rotor (Section 24). The starter clutch is mounted on the back of the alternator rotor.

2   Hold the alternator rotor with one hand and try to rotate the starter driven gear with the other hand **(see illustration)**. It should rotate clockwise smoothly, but not rotate counterclockwise at all.

21.2 Hold the alternator rotor and try to turn the starter driven gear; it should turn smoothly clockwise, but not at all counterclockwise

21.3 Inspect the driven gear and its bearing

21.4a Remove the six Torx bolts . . .

21.4b . . . then remove the outer clutch and the roller assembly

3   If the gear rotates both ways or neither way, or if its movement is rough, remove it and its needle roller bearing from the alternator rotor **(see illustration)**.

4   Remove the Torx bolts and lift off the outer clutch **(see illustrations)**. Inspect the outer clutch and one-way clutch for wear and damage and replace them if any problems are found.

5   Installation is the reverse of the removal steps, with the following additions:

a) *Place the flange side (wide side) of the outer clutch against the alternator rotor.*

b) *Before you tighten the Torx bolts, turn the starter driven gear (see Step 2). It should rotate clockwise only; if it rotates counterclockwise, the one-way clutch is installed backwards.*

c) *Apply non-permanent thread locking agent to the threads of the Torx bolts and tighten them to the torque listed in this Chapter's Specifications.*

Reduction gears

Refer to illustrations 21.7 and 21.8

6   Remove the alternator cover (Section 24). The reduction gears connect the starter motor shaft to the starter clutch on the back of the alternator rotor.

7   If you're working on a 350 model, pull out the reduction gear shaft and remove the gear **(see illustration)**.

8   If you're working on a 250 model, remove the idler gear and shaft from the crankcase **(see illustration)**. Remove the reduction gear, its snap-ring and thrust washer.

9   Check the gears for chipped or worn teeth and replace them if necessary. Replace the thrust washer if it looks worn.

10   Spin the inner race of the starter shaft bearing in the alternator cover. If it's rough, loose or noisy, obtain a new bearing and put it in a freezer so it will shrink. Heat the case around the bearing with a hair dryer or similar tool and tap the bearing out, then tap the new bearing in with a bearing driver or socket the same diameter as the bearing. **Caution:** *Don't heat the cover with a torch or it may warp, which will cause it to seal poorly against the rear crankcase cover.*

11   Installation is the reverse of the removal steps.

22 Charging system testing - general information and precautions

1   If the performance of the charging system is suspect, the system as a whole should be checked first, followed by testing of the individual components (the alternator and the regulator/rectifier). **Note:** *Before beginning the checks, make sure the battery is fully charged and that all system connections are clean and tight.*

2   Checking the output of the charging system and the performance of the various components within the charging system requires the use of a voltmeter, ammeter and ohmmeter or the equivalent multimeter.

3   When making the checks, follow the procedures carefully to prevent incorrect connections or short circuits, as irreparable damage to electrical system components may result if short circuits occur.

4   If the necessary test equipment is not available, it is recommended that charging system tests be left to a dealer service department or a reputable motorcycle repair shop.

23 Charging system - output test

Refer to illustration 23.3

1   If a charging system problem is suspected, perform the following checks. Start by removing the seat for access to the battery

21.7 On 350 models, pull out the reduction gear shaft and remove the gear

21.8 On 250 models, remove the idler gear and shaft, then remove the reduction gear, its snap-ring and thrust washer

23.3 Connect a voltmeter between the battery terminals and measure voltage output with the engine running

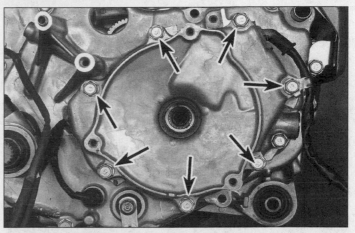

24.6a On 350 models, remove the alternator cover bolts (arrows) - one bolt secures a wiring harness . . .

(see Chapter 7).

2   Start the engine and let it warm up to normal operating temperature.

3   With the engine idling, attach the positive (red) voltmeter lead to the positive (+) battery terminal and the negative (black) lead to the battery negative (-) terminal **(see illustration)**. The voltmeter selector switch (if equipped) must be in the 0-20 DC volt range.

4   Slowly increase the engine speed to 5000 rpm and compare the voltmeter reading to the value listed in this Chapter's Specifications.

5   If the output is as specified, the alternator is functioning properly.

6   Low voltage output may be the result of damaged windings in the alternator stator coils or wiring problems. Make sure all electrical connections are clean and tight, then refer to the following Sections to check the alternator stator coils and the regulator/rectifier.

7   High voltage output (above the specified range) indicates a defective voltage regulator/rectifier. Refer to Section 25 for regulator testing and replacement procedures.

24 Alternator stator coils, rotor and cover(s) - check and replacement

Stator coil check

1   Where necessary for access, remove the seat and left side cover (see Chapter 7).

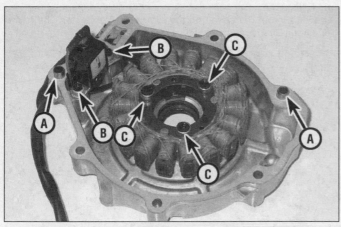

24.6b . . . then pull off the cover, noting the locations of the dowels

A)   Dowels   C)   Stator screws
B)   Pulse generator screws

2   If you're working on a Rancher, disconnect the white five-pin connector in the regulator wiring harness. If you're working on a Recon, disconnect the connector from the regulator/rectifier. If you're working on a TRX250EX, disconnect the three-pin red connector. Connect an ohmmeter between the yellow wire terminals in the side of the connector that runs back to the stator coils on the rear of the engine. The readings should be within the range listed in this Chapter's Specifications. On Rancher and TRX250EX models, replace the stator coils. If the readings are much outside the value listed in this Chapter's Specifications. On Recon models, repeat the test at the three-pin red alternator connector. If the readings are still outside the Specifications replace the stator coils as described below.

3   Connect the ohmmeter between a good ground on the vehicle and each of the connector terminals in turn. The meter should indicate infinite resistance (no continuity). If not, replace the stator coils.

Stator coil replacement

Refer to illustrations 24.6a, 24.6b and 24.8

4   Remove the recoil starter and its pulley (see Chapter 2).

Rancher models

5   Remove the engine from the vehicle (see Chapter 2).

6   Remove the alternator cover bolts, noting that one bolt secures a wiring harness **(see illustration)**. Pull off the cover, noting the locations of the dowels **(see illustration)**.

7   Disconnect the pulse generator connector and remove its Allen bolts **(see illustration 24.6b)**. Unscrew the three Allen bolts that secure the stator to the cover and remove it, together with the pulse generator and wiring harness.

8   Check the cover oil seal for wear or damage **(see illustration)**. It's

24.8 Check the cover oil seal for wear or damage

24.12a On 250 models, remove the alternator cover bolts (arrows)

24.12b . . . then pull off the cover, noting the locations of the dowels (arrows)

a good idea to replace it whenever the cover is removed. To replace the seal pry it out, taking care not to damage the cover or stator. Drive in a new seal using a seal driver or socket the same diameter as the seal.

Recon and TRX250EX models

Refer to illustrations 24.12a, 24.12b, 24.13 and 24.14

9   Disconnect the reverse and neutral switch wires (see Section 12). Follow the pulse generator and stator wiring harnesses from the alter-

24.13 Remove the pulse generator and stator Allen bolts (arrows) and note how the grommet fits in the notch

nator cover to the connectors and disconnect them.
10   Disconnect the reverse control cable and remove the control arm (see Chapter 2).
11   Remove the starter motor (see Section 19).
12   Remove the alternator cover bolts **(see illustration)**. Take the cover off the engine and locate the dowels **(see illustration)**.
13   Disconnect the pulse generator connector and remove its Allen bolts **(see illustration)**. Unscrew the three Allen bolts that secure the stator to the cover and remove it, together with the pulse generator and wiring harness.
14   While the cover is off, check the oil seal for signs of leakage **(see illustration)**. If any problems are found, carefully pry the seal out of the cover, taking care not to damage the seal bore. Drive in a new seal with a seal driver or socket the same diameter as the seal.

All models

Refer to illustration 24.15

15   Installation is the reverse of the removal steps, with the following additions:

   a) *Tighten the stator coil and pulse generator Allen bolts to the torques listed in this Chapter's Specifications.*
   b) *Remove all old gasket material from the alternator cover and crankcase. Use a new gasket on the alternator cover and smear a film of sealant across the wiring harness grommet.*
   c) *Align the pin on the neutral-reverse switch with the slot in the shaft* **(see illustration)**.
   d) *Tighten the cover bolts evenly, in a criss-cross pattern.*
   e) *Don't forget the O-ring on the rotor bolt. Tighten the bolt to the torque listed in this Chapter's Specifications.*

24.14 Check the cover oil seal for wear or damage

24.15 Align the switch pin with this shaft slot

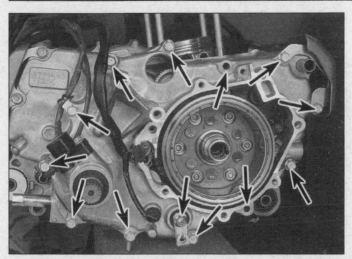

24.19a Remove the rear crankcase cover bolts (arrows) . . .

24.19b . . . take the cover off, locate the dowels (arrows) and remove the washer from the spindle at the bottom of the case

Rear crankcase cover (Rancher models)

Removal

Refer to illustrations 24.19a, 24.19b and 24.20

16   Remove the alternator cover as described above.

17   Remove the reverse lock lever and cable retainer (see Chapter 2).

18   Remove the speed sensor from the rear cover and disconnect the oil temperature sensor wire.

19   Remove the cover bolts **(see illustration)**. Take the cover off, locate the dowels and remove the washer from the spindle at the bottom of the case **(see illustration)**.

20   Check the seals in the cover for wear or damage **(see illustration)**. If any problems are found, carefully pry the seals out of the cover, taking care not to damage the seal bore. Drive in a new seal with a seal driver or socket the same diameter as the seal. The open sides of the seals face into the engine.

Installation

21   Installation is the reverse of the removal steps, with the following additions:

  a) *Make sure the dowels and thrust washer are in position.*

  b)   *Apply a thin coat of non-hardening gasket sealant to the gasket surface of the cover just before installing it. Lubricate the oil seal lips with clean engine oil.*

  c) *Tighten the cover bolts evenly, in a criss-cross pattern. Tighten the bolts securely, but don't overtighten them and strip the threads.*

Rotor

Removal

Refer to illustrations 24.24 and 24.26

Caution: *To remove the alternator rotor, the special Honda puller (part no. 07YMC-HN40100 for Rancher models; 07733-0010001 or 07933-2000000 for Recon and TRX250EX models) or an aftermarket equivalent will be required. Don't try to remove the rotor without the proper puller, as it's almost sure to be damaged. Pullers are readily available from motorcycle dealers and aftermarket tool suppliers.*

22   If you're working on a Rancher, remove the rear crankcase cover as described above.

23   If you're working on a Recon or TRX250EX, remove the starter reduction gears (see Section 21).

24   Thread an alternator puller into the center of the rotor and use it to remove the rotor **(see illustration)**. If the rotor doesn't come off easily, tap sharply on the end of the puller to release the rotor's grip on the tapered crankshaft end.

25   Pull the rotor off, together with the starter clutch.

26   Remove the needle roller bearing and washer from the crankshaft **(see illustration)**.

27   Check the Woodruff key **(see illustration 24.26)**; if it's not secure in its slot, pull it out and set it aside for safekeeping. A convenient method is to stick the Woodruff key to the magnets inside the rotor, but be certain not to forget it's there, as serious damage to the rotor and stator coils will occur if the engine is run with anything stuck to the magnets.

24.20 Check the seals in the cover for wear or damage

24.24 Remove the rotor with a puller designed for the purpose

24.26 Remove the needle roller bearing and washer from the crankshaft

24.32 Temporarily bolt the recoil starter pulley to the crankshaft to seat the rotor

24.33 Be sure there aren't any small metal objects stuck to the rotor magnets; an inconspicuous item like this Woodruff key (arrow) can ruin the rotor and stator if the engine is run

Installation

Refer to illustrations 24.32 and 24.33

28   Degrease the center of the rotor and the end of the crankshaft.

29   Make sure the Woodruff key is positioned securely in its slot. Lubricate the needle roller bearing with clean engine oil and install the washer and needle roller bearing on the crankshaft **(see illustration 24.26)**.

30   Check the make sure the starter reduction shaft is installed in the engine.

31   Align the rotor slot with the Woodruff key. Place the rotor, together with the starter clutch and starter driven gear, on the crankshaft.

32   **Note:** *Don't skip this step, or the rotor magnets may pull the rotor off the crankshaft when the alternator cover is installed.* Temporarily install the recoil starter pulley and tighten its bolt to seat the rotor securely on the crankshaft **(see illustration)**. Remove the bolt and pulley once this is done.

33   Take a look to make sure there isn't anything stuck to the inside of the rotor **(see illustration)**.

34   The remainder of installation is the reverse of the removal steps.

25   Regulator/rectifier - check and replacement

Check

Refer to illustration 25.1

1   The regulator/rectifier on 350 models is mounted on the right side of the vehicle frame under the rear fender **(see illustration)**. On 250 models, it's on the left side.

2   Disconnect the wiring harness connector from the regulator/ rectifier.

3   Connect a voltmeter between the red wire terminal in the harness side of the connector and a good body ground. It should indicate battery voltage (approximately 13 volts).

4   Connect an ohmmeter between the green wire terminals in the harness side of the connector and a good body ground. The ohmmeter should indicate continuity.

5   Connect the ohmmeter between the yellow wire terminals in the harness side of the connector. The ohmmeter should indicate 0.1 to 1.0 ohms.

5   If the voltage or ohmmeter readings aren't as described, check the wiring for breaks or poor connections. If there's a problem in Step 5, also check the alternator stator coils as described in Section 24.

6   If the wiring harness and stator coils are good, the regulator/rectifier may be defective (especially if the charging system output test in Section 23 indicates excessive output). It's a good idea to have it tested by a Honda dealer or substitute a known good unit before condemning it as trash.

26   Electric shift system (TE/FE models) - check and component replacement

1   This system replaces the foot shift pedal on the external shift linkage with an electric motor, controlled by a switch on the left handlebar.

Troubleshooting

1   Switch the ignition Off and back to On. The system may reset and work properly, depending on the problem. If it does, operate the vehicle; if the problem occurs again, try to note the conditions that led up to the failure.

2   Check the main fuse, motor fuse, light fuse and ignition fuse and replace them if necessary (see Section 5). If you find a blown fuse, check its circuit for a short or bad connection.

3   Try to shift the transmission through the gear positions, using the removable hand lever supplied in the vehicle's tool kit. If you can't shift through the gears using the tool, there's a problem in the mechanical shift linkage (internal or external) (see Chapter 2).

4   Make sure the battery is in good condition and fully charged (see Chapter 1 and Section 4).

5   Check the system wiring for breaks or bad connections, referring to the wiring diagrams at the end of the manual.

6   Check the clutch adjustment (see Chapter 1).

7   If you're working on a 2000 through 2002 350 model, the electric shift system will stop working if the electronic control unit detects a component failure. If the problem is with second, third or fourth gear shifting (which is controlled by the ECU), the instrument cluster will dis-

25.1 The regulator/rectifier unit is mounted on the frame under the rear fender (arrow) (350 shown; 250 similar)

play dashes, rather than indicating any of these gear positions.

8   2003 and later 350 models, and all 250 models, can isolate the problem to a specific circuit or component by flashing trouble codes, using the neutral indicator in the instrument cluster.

Reading trouble codes (2003-on 350 and all 250 models)

9   Turn the ignition On (if it isn't already), then Off and back to On. If there's a problem in the system, it will be indicated by the gear position indicator, which will flash a number of times to indicate a trouble code (one flash equals code 1, two flashes equal code 2, etc.). Trouble codes and their specific problem areas are:

> *Code 0 (no flashes) - The system is functioning normally.*
> *Code 1 (one flash) - Electronic control unit*
> *Code 2 (two flashes) - Shift switch or its circuit*
> *Code 3 (three flashes) - Angle sensor or its circuit*
> *Code 4 (four flashes) - Gear position switch or its circuit*
> *Code 5 (five flashes) - ECU motor circuit (defective ECU)*
> *Code 6 (six flashes) - ECU failsafe relay circuit (defective ECU)*
> *Code 7 (seven flashes) - ECU voltage converter circuit (defective ECU)*
> *Code 8 (eight flashes) - Angle sensor, control motor, ECU or circuit*
> *Code 9 (nine flashes) - Angle sensor, its circuit or ECU*
> *Code 10 (10 flashes) - Pulse generator or its circuit*
> *Code 11 (11 flashes) - Speed sensor, its circuit or ECU*
> *Code 12 (12 flashes) - Gear position switch, its circuit or ECU*

Angle sensor check and replacement

Check

10   With the ignition switch in the Off position and the transmission in Neutral, disconnect the 22-pin electrical connector from the electronic control unit.

11   Locate the terminal pins for the black/red and blue/green wires in the sensor wiring harness. Connect an ohmmeter between the terminals. The reading (angle sensor neutral resistance) should be within the values listed in this Chapter's Specifications. Write the reading down for use in the next Steps.

12   Move the ohmmeter to the blue/green and yellow/blue terminals. Using the manual shift handle supplied in the vehicle's tool kit, shift the transmission up through the gears.

Rancher models

13   Note the ohmmeter reading and divide it by the reading written down in Step 11. The result should be less than 0.4 ohms.

14   With the ohmmeter still connected to the blue/green and yellow/red terminals, use the manual shift handle to shift the transmission

26.18 Unbolt the cover from the angle sensor (Rancher shown; Recon similar) . . .

down through the gears. Note the ohmmeter reading and divide it by the reading written down in Step 12. The result should be greater than 0.6 ohms.

Recon models

15   The ohmmeter reading should range from zero to 5000 ohms as you shift through the gears.

All models

16   If the readings aren't within the specified range, remove the cover from the angle sensor, then repeat the measurements in Steps 11 through 15 at the angle sensor terminals (the sensor terminals, not the harness terminals). If the measurements are now OK, check the wiring for breaks or bad connections. If they are still not OK, replace the angle sensor.

17   Reconnect the 22-pin connector to the ECU, then connect a voltmeter between the black/red and blue/green terminals at the angle sensor end of the wiring harness. With the key turned to On, the voltmeter should indicate 4.7 to 5.3 volts. If not, follow the harness back to the ECU, checking for breaks or bad connections.

Replacement

Refer to illustrations 26.18, 26.19a, 26.19b, 26.19c and 26.20

18   Remove the cover from the shift control motor and angle sensor **(see illustration)**.

19   Disconnect the electrical connector from the angle sensor. Remove the Allen bolts, take the angle sensor off the engine and remove its O-ring **(see illustrations)**.

26.19a Note which way the angle sensor points, then remove its Allen bolts (lower arrows); DO NOT remove the shift motor Allen bolts (this is a Rancher)

26.19b Here's the Recon angle sensor; note which way it points and remove its Allen bolts . . .

26.19c . . . on all models, remove the angle sensor and its O-ring

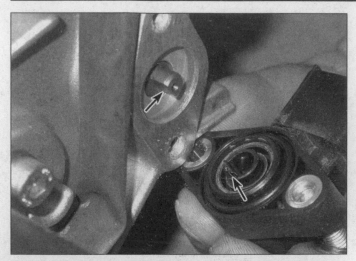

26.20 Check the flats on the spindle and in the motor (arrows) for wear; align the flats on installation

26.23 Here are the Recon shift motor mounting bolts (arrows); DO NOT remove the Allen bolts

20   As a final check, inspect the drive spindle on the engine and the hole in the angle sensor that the spindle fits into **(see illustration)**. Make sure they're not worn or damaged. Connect an ohmmeter between the angle sensor terminals that connect to the blue/green and yellow/blue wires. Slowly rotate the angle sensor shaft, using a screwdriver or similar tool inserted into the shaft hole in the angle sensor. When the shaft is turned counterclockwise, resistance should increase (steadily, not in jumps). Resistance should decrease steadily when the shaft is turned clockwise.

21   Installation is the reverse of the removal steps. Install a new O-ring and be sure to align the flat in the angle sensor with the flat on the gearshift spindle **(see illustrations 26.19c and 26.20)**.

Shift control motor check and replacement

Check

Refer to illustration 26.23

22   Remove the angle sensor as described above.

23   Follow the wiring harness from the shift control motor to the connector and disconnect it. Remove the motor's mounting bolts and take it off the engine **(see illustration 26.18 or the accompanying illustration)**. **Caution:** *Remove the hex-head mounting bolts only. DO NOT remove the Allen bolts, or the motor housing will separate and the brushes will fall out of position.*

24   Use a pair of jumper wires with probes on the end to connect the control motor directly to a 12-volt battery (the vehicle's battery will do if it's fully charged). The motor should turn. If it doesn't, replace it.

25   Installation is the reverse of the removal steps. Use a new O-ring, coated with clean engine oil.

Reduction gear check and replacement

Refer to illustrations 26.28, 26.29, 26.30a and 26.30b

26   Remove the motor as described above.

27   If you're working on a 350 model, unbolt the gear cover and take it off the engine. If you're working on a 250 model, the motor and gear cover come off together.

28   Pull out the lower gear, upper gear and middle gear **(see illustration)**. Remove the O-ring from its groove.

29   Check all parts, especially gear teeth, for wear and damage. Replace the gears as a set if problems are found. Check the bearings in the cover for rough, loose or noisy movement **(see illustration)**.

30   Installation is the reverse of the removal steps. Lubricate the gear's pivot holes with Esso Unirex N2 or N3 grease or equivalent **(see illustration)**. Apply the same grease to the gear teeth. When you install the lower gear, align its punch mark with the wide spline on the shaft **(see illustration)**.

26.28 Remove the reduction gears and O-ring (Rancher shown; Recon similar)

26.29 Check the bearings inside the cover

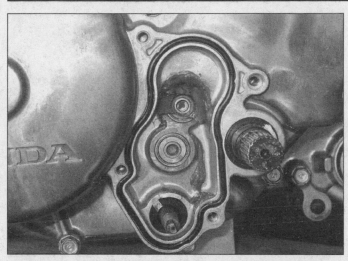

26.30a Lubricate the reduction gear pivot bores and gears with special grease

26.30b Align the lower gear punch mark with the wide spline

Shift switch

31   The shift switch is located on the left handlebar. To remove and install the switch, refer to Section 14. To test switch continuity, refer to the continuity diagram, which is included in the wiring diagram at the end of this manual.

Gear position switch

32   This procedure applies to the gear position switch mounted inside the rear crankcase cover on Rancher models and inside the alternator cover on 2004 and earlier electric shift Recon models, all 2005 and later Recon models and 2006 and later TRX250EX models. The switch indicates neutral and reverse on foot shift models, and all gear positions on electric shift models.

Check

33   Disconnect the switch electrical connector (it can be identified by its wire colors).
34   Connect the negative terminal of an ohmmeter to ground. Connect the positive terminal to each of the gear position terminals in turn, then shift gears (use the hand shift tool on electric shift models).
35   On foot shift models, there should be continuity as follows:

 a) *In neutral: Light green/red wire's terminal and ground*
 b) *In reverse: Gray wire's terminal and ground*

36   On electric shift models, there should be continuity as follows:

 a) *In neutral: Light green/red wire's terminal and ground*
 b) *In reverse: Gray wire's terminal and ground*
 c) *In first gear: White/green wire's terminal and ground*
 d) *In second gear: White/red wire's terminal and ground*
 e) *In third gear: Blue wire's terminal and ground*
 f) *In fourth gear: Yellow wire's terminal and ground*
 g) *In fifth gear: Light blue/white wire's terminal and ground*

37   If the switch doesn't perform correctly, replace it.

Replacement

Refer to illustrations 26.40 and 26.41
38   Remove the alternator cover (see Section 24). If you're working on a Rancher, remove the rear crankcase cover as well.
39   If you're working on an electric shift model, disconnect the wiring connector for the reverse switch.
40   Work the grommet out of its slot in the crankcase **(see illustration)**.
41   Unbolt the switch and take it out **(see illustration)**.
42   Installation is the reverse of the removal Steps. Align the switch

26.40 Work the grommet (arrow) out of its slot in the crankcase

26.41 Unbolt the switch and take it out

pin with the slot in shaft **(see illustration 24.15)**. Use non-permanent thread locking agent on the threads of the switch bolt and tighten it to the torque listed in this Chapter's Specifications.

27.8 Disconnect the electrical connector from the speed sensor (arrow)

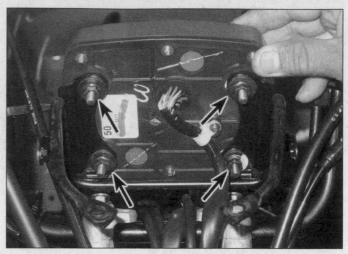

27.17 Remove the cluster mounting nuts (arrows) and take the cluster off

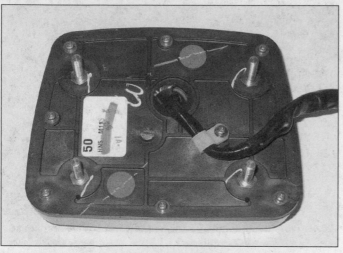

27.18a Remove the screws from the back of the case

27 Instrument cluster - check, removal and installation

Wiring check

1 Disconnect the cluster's 14-pin electrical connector. You may need to dismount the ignition switch and handlebar switch connectors from the frame for access.

2 Connect a voltmeter between ground and the red wire's terminal in the harness side of the connector. There should be battery voltage (approximately 12 volts) whether the key is On of Off.

3 Connect the voltmeter between ground and the black/brown wire's terminal in the harness side of the connector. There should be battery voltage with the key On.

4 Connect an ohmmeter between ground and the green wire's terminal in the harness side of the connector. There should be continuity.

5 Remove the right side cover for access (see Chapter 7). Disconnect the speed sensor's connector (white, three pins). Connect the voltmeter between the green and black/blue wire terminals in the harness side of the connector. There should be battery voltage with the key On.

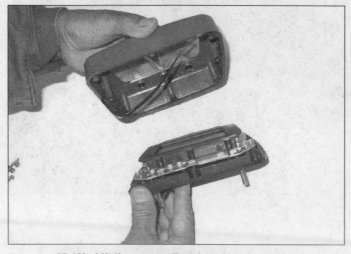

27,18b Lift the cover off and remove the gasket

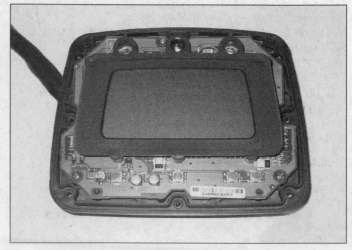

27.19 Remove the screws to separate the circuit board from the panel

6   If any of the above tests do not produce the correct result, check the related wire for breaks or bad connections.

Speed sensor check

Refer to illustration 27.8

7   Jack up the vehicle so all four wheels are off the ground and support it securely on jackstands.

8   Disconnect the electrical connector from the speed sensor **(see illustration)**. Connect a voltmeter between the terminals for the green wire and pink/green wire in the speed sensor side of the connector.

9   Turn the key to On, but don't start the engine. Turn the rear wheels by hand and watch the voltmeter. It should pulse regularly between zero and five volts. If not, replace the sensor. If it performs correctly, check the wiring for breaks or bad connections. If the wiring is good, the instrument cluster may be at fault. Since this can't be returned once purchased, it's a good idea to have it tested by a dealer service department or other qualified shop.

Speed sensor replacement

10   If you haven't already done so, remove the right side cover for access (see Chapter 7).

11   Disconnect the speed sensor electrical connector.

12   Remove the sensor mounting bolts **(see illustration 27.8)**. Remove the sensor, its spacer and two O-rings (one on each side of the spacer).

13   Installation is the reverse of the removal Steps. Use new O-rings, coated with clean engine oil. Tighten the bolts securely, but don't overtighten them and strip the threads.

Instrument cluster replacement

Refer to illustrations 27.17, 27.18a, 27.18b and 27.19

14   Remove the seat and front fender (see Chapter 7).

15   Disconnect the negative cable from the battery.

16   Follow the wiring harness from the cluster to the connector and disconnect it.

17   Remove the cluster mounting nuts and lift it out **(see illustration)**.

18   Remove the case screws and separate the case and gasket from the cluster **(see illustrations)**.

19   Remove the panel screws and separate the instrument panel from the case **(see illustration)**.

20   Installation is the reverse of the removal Steps.

28 Wiring diagrams

Prior to troubleshooting a circuit, check the fuses to make sure they're in good condition. Make sure the battery is fully charged and check the cable connections.

When checking a circuit, make sure all connectors are clean, with no broken or loose terminals or wires. When unplugging a connector, don't pull on the wires - pull only on the connector housings themselves.

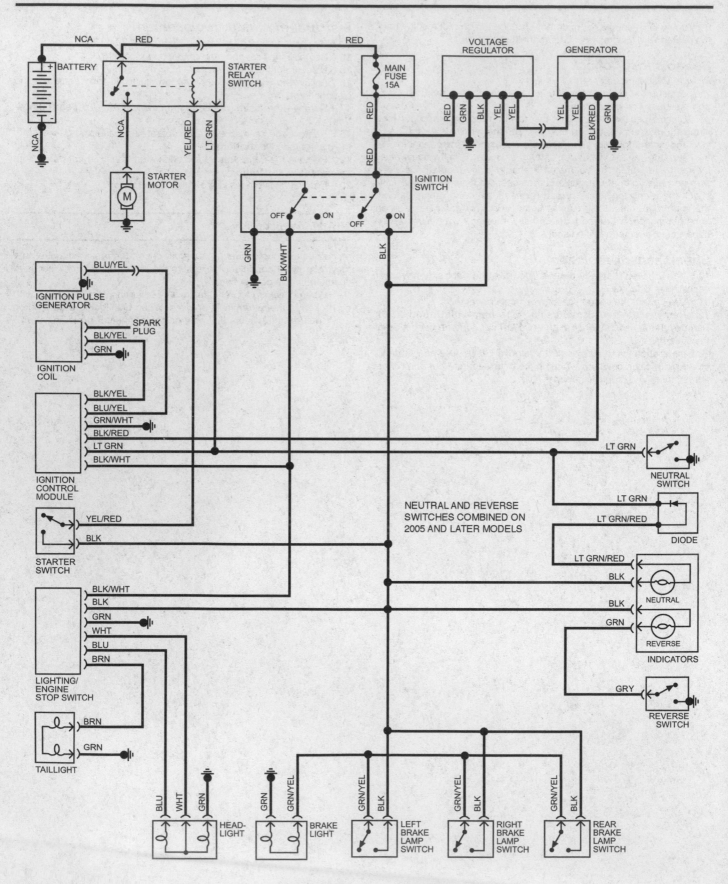

Foot shift Recon and TRX250EX

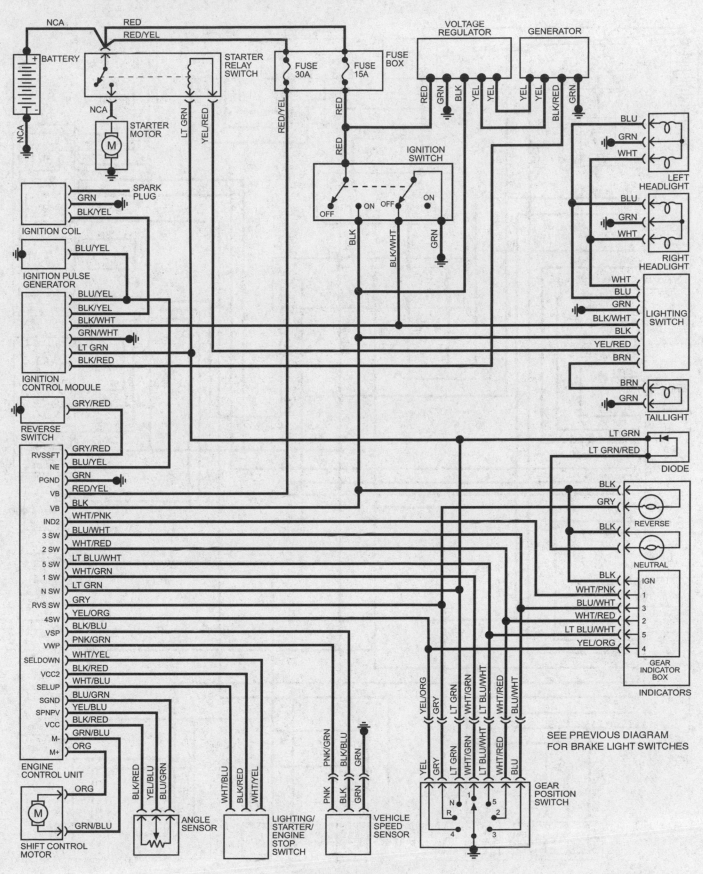

Electric shift Recon

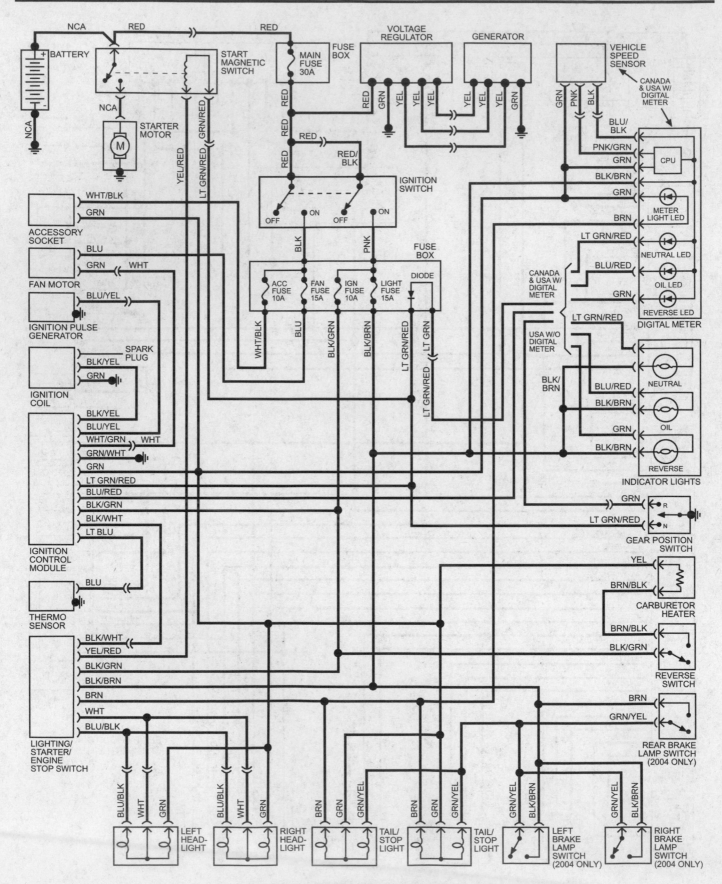

2000 and later Rancher TRX350 TM & FM models

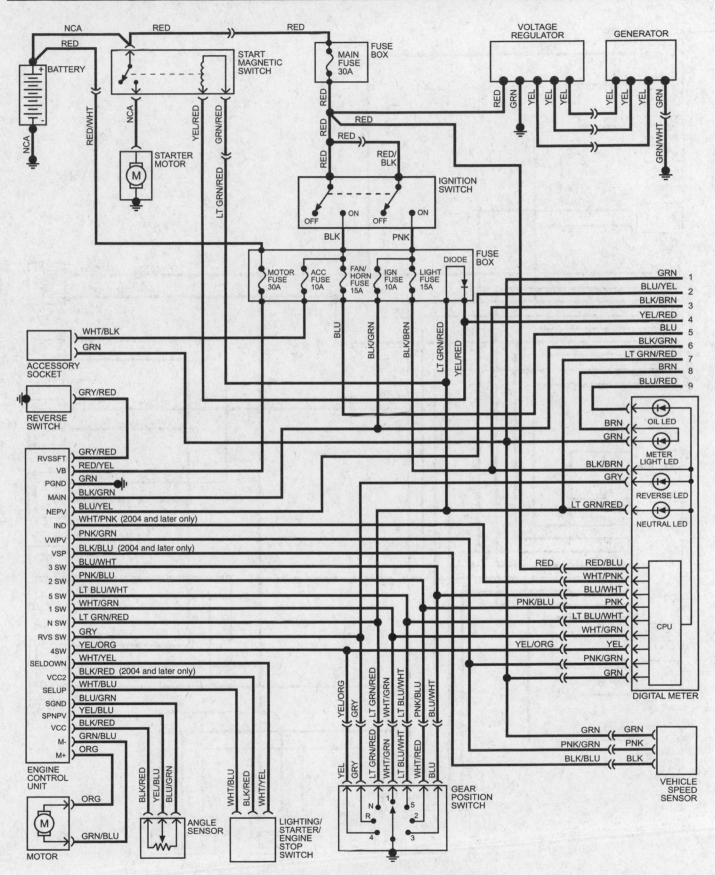

2000 and later Rancher TRX350 TE & FE models (1 of 2)

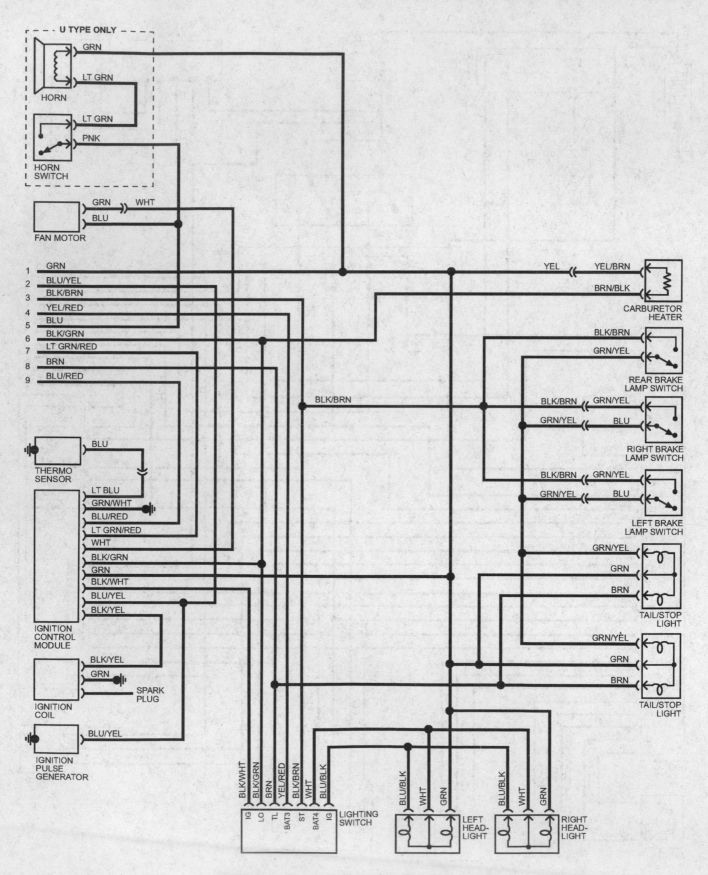

2000 and later Rancher TRX350 TE & FE models (2 of 2)

Index

A

About this manual, 0-5
Acknowledgements, 0-2
Air cleaner housing, removal and installation, 3-12
Air cleaner, filter element and drain tube cleaning, 1-14
Alternator stator coils, rotor and cover(s), check and replacement, 8-15
ATV chemicals and lubricants, 0-17
Axle shaft and housing, rear, removal, inspection and installation, 5-12

B

Battery
  charging, 8-3
  check, 1-6
  inspection and maintenance, 8-3
Bodywork and frame, 7-1 through 7-6
  bumper, front, removal and installation, 7-3
  cargo rack, removal and installation
    front, 7-3
    rear, 7-5
  fender, removal and installation
    front, 7-5
    rear, 7-5
  footpegs and mudguards, removal and installation, 7-4
  frame, general information, inspection and repair, 7-6
  fuel tank cover, removal and installation, 7-3
  seat, removal and installation, 7-2
  side covers, removal and installation, 7-2
Brake
  bleeding, 6-8
  caliper (TRX250EX models), removal, overhaul and installation, 6-15
  disc and hub (TRX250EX), inspection, removal and installation, 6-17
  drum
    inspection, waterproof seal replacement and 2WD bearing replacement, 6-3
    removal and installation
      front, 6-2
      rear, 6-9
  hoses and lines, inspection and replacement, 6-9
  lever and pedal freeplay, check and adjustment, 1-8
  light switches, check and replacement, 8-6
  master cylinder, removal, overhaul and installation, 6-7
  pads, front (TRX250EX models), replacement, 6-14
  panel, removal, inspection and installation
    front, 6-7
    rear, 6-11
  pedal, rear brake lever and cables, removal and installation, 6-13
  shoes, removal, inspection and installation
    front, 6-4
    rear, 6-11
  system, general check, 1-7
  wheel cylinders, removal, overhaul and installation, 6-5
Brake light and taillight bulbs, replacement, 8-6
Brakes, wheels and tires, 6-1 through 6-22
Buying parts, 0-7

C

Caliper, front brake (TRX250EX models), removal, overhaul and installation, 6-15
Cam chain tensioner, removal and installation, 2-39
Camshaft, chain and sprockets, removal, inspection and installation, 2-40
Carburetor
  disassembly, cleaning and inspection, 3-6
  heater (350 models), removal and installation, 3-15
  overhaul, general information, 3-4
  reassembly and float height check, 3-11
  removal and installation, 3-4
Cargo rack, removal and installation
  front, 7-3
  rear, 7-5
Centrifugal clutch, removal, inspection and installation, 2-25
Change clutch and release mechanism, removal, inspection and installation, 2-29
Charging system
  alternator stator coils, rotor and cover(s), check and replacement, 8-15
  output test, 8-14
  regulator/rectifier, check and replacement, 8-18
  testing, general information and precautions, 8-14
Chemicals and lubricants, 0-17
Choke
  cable, removal and installation, 3-14
  operation check, 1-12
Clutch, check and freeplay adjustment, 1-10
Coil, ignition, check, removal and installation, 4-2
Conversion factors, 0-18
Crankcase
  components, inspection and servicing, 2-51
  disassembly and reassembly, 2-49
Crankshaft and balancer, removal, inspection and installation, 2-57
Cylinder compression, check, 2-10
Cylinder head and valves, disassembly, inspection and reassembly, 2-17
Cylinder, removal, inspection and installation, 2-20

D

Diagnosis, 0-20
Differential and driveshaft, front, removal, inspection and installation, 5-11
Differential oil, change, 1-14
Differential, rear, removal, inspection and installation, 5-13
Disc and hub, front brake (TRX250EX), inspection, removal and installation, 6-17
Driveaxle, front (4WD models)
    boot
        check, 1-11
        replacement and CV joint overhaul, 5-9
    removal and installation, 5-8
Driveshaft, rear, removal, inspection and installation, 5-16

E

Electrical system, 8-1 through 8-28
    alternator stator coils, rotor and cover(s), check and replacement, 8-15
    battery
        charging, 8-3
        inspection and maintenance, 8-3
    brake light switches, check and replacement, 8-6
    charging system
        output test, 8-14
        testing, general information and precautions, 8-14
    electric shift system (TE/FE models), check and component replacement, 8-18
    electrical troubleshooting, 8-2
    fuses, check and replacement, 8-4
    handlebar switches
        check, 8-9
        removal and installation, 8-9
    headlight
        aim, check and adjustment, 8-6
        bulb replacement, 8-5
    ignition main (key) switch, check and replacement, 8-9
    indicator bulbs, replacement, 8-7
    instrument cluster, check, removal and installation, 8-22
    lighting system, check, 8-5
    neutral and reverse switches, check and replacement, 8-9
    oil cooling system, check and component replacement, 8-7
    regulator/rectifier, check and replacement, 8-18
    starter
        clutch and reduction gears, removal, inspection and installation, 8-13
        motor
            disassembly, inspection and reassembly, 8-11
            removal and installation, 8-10
        relay, check and replacement, 8-10
        switch, check and replacement, 8-10
    taillight and brake light bulbs, replacement, 8-6
    wiring diagrams, 8-23
Engine oil/filter, change, 1-12
Engine, clutch and transmission, 2-1 through 2-62
    cam chain tensioner, removal and installation, 2-39
    camshaft, chain and sprockets, removal, inspection and installation, 2-40
    centrifugal clutch, removal, inspection and installation, 2-25

    change clutch and release mechanism, removal, inspection and installation, 2-29
    crankcase
        inspection and servicing, 2-51
        disassembly and reassembly, 2-49
    crankshaft and balancer, removal, inspection and installation, 2-57
    cylinder compression, check, 2-10
    cylinder head and valves, disassembly, inspection and reassembly, 2-17
    cylinder, removal, inspection and installation, 2-20
    engine
        disassembly and reassembly, general information, 2-12
        removal and installation, 2-11
    external shift mechanism, removal, inspection and installation, 2-44
    initial start-up after overhaul, 2-59
    major engine repair, general note, 2-10
    oil pump and cooler, removal, inspection and installation, 2-35
    operations possible with the engine in the frame, 2-10
    operations requiring engine removal, 2-10
    piston rings, installation, 2-24
    piston, removal, inspection and installation, 2-22
    primary drive gear, removal, inspection and installation, 2-35
    recoil starter, removal, inspection and installation, 2-43
    recommended break-in procedure, 2-59
    reverse lock mechanism, cable replacement, removal, inspection and installation, 2-42
    rocker arms, pushrods and cylinder head, removal, inspection and installation, 2-13
    transmission shafts and shift drum, removal, inspection and installation, 2-52
    valve cover (1997 through 2001 Recon and all 350 models), removal, inspection and installation, 2-13
    valve lifters, removal, inspection and installation, 2-25
    valves/valve seats/valve guides, servicing, 2-16
Exhaust system
    inspection and spark arrester cleaning, 1-16
    removal and installation, 3-14
External shift mechanism, removal, inspection and installation, 2-44

F

Fasteners, check, 1-18
Fault finding, 0-20
Fender, removal and installation
    front, 7-5
    rear, 7-5
Fluid levels, check, 1-5
Fluids and lubricants, recommended, 1-3
Footpegs and mudguards, removal and installation, 7-4
Fraction/decimal/millimeter equivalents, 0-19
Frame, general information, inspection and repair, 7-6
Front brake
    caliper (TRX250EX models), removal, overhaul and installation, 6-15
    disc and hub (TRX250EX), inspection, removal and installation, 6-17

drums
   inspection, waterproof seal replacement and 2WD
      bearing replacement, 6-3
   removal and installation, 6-2
master cylinder, removal, overhaul and installation, 6-7
pads (TRX250EX models), replacement, 6-14
panel, removal, inspection and installation, 6-7
shoes, removal, inspection and installation, 6-4
Front cargo rack and bumper, removal and installation, 7-3
Front differential and driveshaft, removal, inspection and installation, 5-11
Front driveaxle (4WD models)
   boot replacement and CV joint overhaul, 5-9
   removal and installation, 5-8
Front fender, removal and installation, 7-5
Front wheel bearings (TRX250EX models), removal, inspection and installation, 6-18
Front wheel cylinders, removal, overhaul and installation, 6-5
Fuel and exhaust systems, 3-1 through 3-16
   air cleaner housing, removal and installation, 3-12
   carburetor
      disassembly, cleaning and inspection, 3-6
      heater (350 models), removal and installation, 3-15
      overhaul, general information, 3-4
      reassembly and float height check, 3-11
      removal and installation, 3-4
   check and filter cleaning, 1-15
   choke cable, removal and installation, 3-14
   exhaust system, removal and installation, 3-14
   idle fuel/air mixture adjustment, 3-3
   tank
      cleaning and repair, 3-3
      cover, removal and installation, 7-3
      removal and installation, 3-2
   throttle cable and housing, removal, installation and adjustment, 3-12
Fuses, check and replacement, 8-4

G
General specifications, 0-8

H
Handlebar switches
   check, 8-9
   removal and installation, 8-9
Handlebars, removal, inspection and installation, 5-2
Headlight
   aim, check and adjustment, 8-6
   bulb replacement, 8-5
Hubs, rear wheel, removal and installation, 6-20

I
Identification numbers, 0-6
Idle fuel/air mixture adjustment, 3-3
Idle speed, check and adjustment, 1-18

Ignition system, 4-1 through 4-4
   coil, check, removal and installation, 4-2
   control module (ICM) , harness check, removal and installation, 4-4
   main (key) switch, check and replacement, 8-9
   pulse generator, check, removal and installation, 4-3
   system, check, 4-1
   timing, general information and check, 4-4
Indicator bulbs, replacement, 8-7
Initial start-up after overhaul, 2-59
Instrument cluster, check, removal and installation, 8-22
Introduction to the Honda Rancher, Recon and Sportrax, 0-5
Introduction to tune-up and routine maintenance, 1-5

L
Lighting system, check, 8-5
Lubricants and chemicals, 0-17
Lubricants and fluids, recommended, 1-3
Lubrication, general, 1-12

M
Maintenance schedule, 1-4
Maintenance techniques, tools and working facilities, 0-10
Major engine repair, general note, 2-10
Master cylinder, brake, removal, overhaul and installation, 6-7

N
Neutral and reverse switches, check and replacement, 8-9

O
Oil cooling system, check and component replacement, 8-7
Oil pump and cooler, removal, inspection and installation, 2-35
Operations possible with the engine in the frame, 2-10
Operations requiring engine removal, 2-10

P
Parts, replacement, buying, 0-7
Piston rings, installation, 2-24
Piston, removal, inspection and installation, 2-22
Primary drive gear, removal, inspection and installation, 2-35
Pulse generator, check, removal and installation, 4-3

R
Rear axle shaft and housing, removal, inspection and installation, 5-12

Rear brake
drum, removal, inspection and installation, 6-9
panel, removal, inspection and installation, 6-11
shoes, removal, inspection and installation, 6-11
Rear cargo rack, removal and installation, 7-5
Rear differential, removal, inspection and installation, 5-13
Rear driveshaft, removal, inspection and installation, 5-16
Rear fender, removal and installation, 7-5
Rear wheel hubs, removal and installation, 6-20
Recoil starter, removal, inspection and installation, 2-43
Recommended break-in procedure, 2-59
Recommended lubricants and fluids, 1-3
Regulator/rectifier, check and replacement, 8-18
Reverse and neutral switches, check and replacement, 8-9
Reverse lock mechanism, cable replacement, removal, inspection and installation, 2-42
Reverse lock system, check and adjustment, 1-9
Rocker arms, pushrods and cylinder head, removal, inspection and installation, 2-13
Routine maintenance and tune-up, introduction, 1-5
Routine maintenance schedule, 1-4

S

Safety first, 0-16
Scheduled maintenance, 1-4
Seat, removal and installation, 7-2
Shock absorbers, removal and installation, 5-4
Side covers, removal and installation, 7-2
Spark plug, replacement, 1-16
Specifications, general, 0-8
Starter
clutch and reduction gears, removal, inspection and installation, 8-13
motor
disassembly, inspection and reassembly, 8-11
removal and installation, 8-10
relay, check and replacement, 8-10
switch, check and replacement, 8-10
Steering system, inspection and toe-in adjustment, 1-19
Steering, suspension and final drive, 5-1 through 5-18
front differential and driveshaft, removal, inspection and installation, 5-11
front driveaxle (4WD models)
boot replacement and CV joint overhaul, 5-9
removal and installation, 5-8
handlebars, removal, inspection and installation, 5-2
rear axle shaft and housing, removal, inspection and installation, 5-12
rear differential, removal, inspection and installation, 5-13
rear driveshaft, removal, inspection and installation, 5-16
shock absorbers, removal and installation, 5-4
steering knuckle
bearing replacement, 5-7
removal, inspection, and installation, 5-6

steering shaft, removal, inspection, bearing replacement and installation, 5-3
suspension arms, removal, inspection, ball-joint replacement and installation, 5-7
swingarm
bearings, check, 5-14
bearings, replacement, 5-16
removal and installation, 5-14
tie-rods, removal, inspection and installation, 5-5
Suspension arms, removal, inspection, ball-joint replacement and installation, 5-7
Suspension, check, 1-19
Swingarm
bearings, check, 5-14
bearings, replacement, 5-16
removal and installation, 5-14

T

Taillight and brake light bulbs, replacement, 8-6
Throttle cable and housing, removal, installation and adjustment, 3-12
Throttle operation/grip freeplay, check and adjustment, 1-11
Tie-rods, removal, inspection and installation, 5-5
Timing, ignition, general information and check, 4-4
Tires, general information, 6-20
Tires/wheels, general check, 1-9
Tools and working facilities, 0-10
Transmission shafts and shift drum, removal, inspection and installation, 2-52
Troubleshooting, 0-20
Tune-up and routine maintenance, 1-1 through 1-20
Tune-up and routine maintenance, introduction, 1-5

V

Valve clearances, check and adjustment, 1-17
Valve cover (1997 through 2001 Recon and all 350 models), removal, inspection and installation, 2-13
Valve lifters, removal, inspection and installation, 2-25
Valves/valve seats/valve guides, servicing, 2-16

W

Wheel bearings, front (TRX250EX models), removal, inspection and installation, 6-18
Wheel cylinders, removal, overhaul and installation, 6-5
Wheels, inspection, removal and installation, 6-19
Wiring diagrams, 8-23
Working facilities, 0-10